Paper Inventions

Machines That Move, Drawings That Light Up, and Wearables and Structures You Can Cut, Fold, and Roll

Kathy Ceceri

MAKER MEDIA

SAN FRANCISCO, CA

Make: Paper Inventions

by Kathy Ceceri

Copyright © 2015 Kathy Ceceri. All rights reserved.

Printed in the United States of America.

Published by Maker Media, Inc., 1005 Gravenstein Highway North, Sebastopol, CA 95472.

Maker Media books may be purchased for educational, business, or sales promotional use. Online editions are also available for most titles (*http://safaribooksonline.com*). For more information, contact our corporate/institutional sales department: 800-998-9938 or corporate@oreilly.com.

Editor: Patrick Di Justo	**Indexer:** Stephen Ingle, WordCo
Production Editor: Shiny Kalapurakkel	**Interior Designer:** David Futato
Copyeditor: Amanda Kersey	**Cover Designers:** Jean Tashima & Sergio Burgos
Proofreader: James Fraleigh	**Illustrator:** Rebecca Demarest

September 2015: First Edition

Revision History for the First Edition
2015-09-01: First Release

See *http://oreilly.com/catalog/errata.csp?isbn=9781457187520* for release details.

978-1-457-18752-0

[LSI]

Table of Contents

Preface

Paper is incredible stuff. It's cheap, easy to use, and easy to recycle. It's lightweight and easy to cut or tear, but incredibly strong when folded, layered, or rolled. It can stand stiff as a board, pop up like a spring, or float like a leaf. It's disposable, but can last for centuries. Its surface can be rough or creamy, smooth or shiny. Sometimes it's so thin you can see through it; other times, it's thick enough to hold globs of paint. But it can also be beautiful all on its own. With so many kinds of paper around—from fancy artists' watercolor paper to slick recycled magazine page to plain old copy paper—it's not surprising that people today use paper to make everything from jewelry to robots. And you can too!

The projects in *Paper Inventions* were inspired by the cool ways you can use paper to explore STEAM—Science, Technology, Engineering, Art, and Math. As you'll discover, there's a lot of overlap between science and math, math and engineering, and even art and technology. That's what makes inventing with crafts materials so much fun!

In each chapter, you'll find projects for beginners as well as experienced Makers. Some can be done in an hour or two. The most complicated may take a few afternoons. Most require only ordinary art and household supplies. But you'll also get suggestions for making them bigger and better using kid-friendly products and materials. You'll also learn where they came from and how they work. You'll get a list of suggested materials and where to get them. And simple step-by-step instructions and illustrations will make it easy to try the projects on your own or with friends and classmates.

Whether you like to make crafts or play with electronics, there's a project that will challenge you and unleash your creativity. Of course, it's important to remember that all inventors run into snags. When your project or new idea doesn't work out the way you planned, don't give up! Learning how to *troubleshoot* by going back over every step until you find the problem is all part of the creative process. Luckily, when you're working with paper, it's not hard to repair a broken piece with a little tape, or even start over if you have to.

Are you ready to give *Paper Inventions* a try? Great! Let's get started!

Acknowledgments

The projects in this book were inspired by many artists who share their work online and in print. Be sure to check out the resources listed at the end of each project. I'd especially like to acknowledge Jie Qi, creator of Circuit Stickers (*http://chibitronics.com/*), whose work on paper circuits has been very helpful. I'd also like to thank the Schuylerville Public Library and students in the Schuylerville High School Life Skills class in New York, who helped with the newspaper dome project.

Conventions Used in This Book

This element signifies a general note, tip, or suggestion.

This element indicates a warning or caution.

Safari® Books Online

Safari Books Online is an on-demand digital library that delivers expert content in both book and video form from the world's leading authors in technology and business.

Technology professionals, software developers, web designers, and business and creative professionals use Safari Books Online as their primary resource for research, problem solving, learning, and certification training.

Safari Books Online offers a range of plans and pricing for enterprise, government, education, and individuals.

Members have access to thousands of books, training videos, and prepublication manuscripts in one fully searchable database from publishers like Maker Media, O'Reilly Media, Prentice Hall Professional, Addison-Wesley Professional, Microsoft Press, Sams, Que, Peachpit Press, Focal Press, Cisco Press, John Wiley & Sons, Syngress, Morgan Kaufmann, IBM Redbooks, Packt, Adobe Press, FT Press, Apress,

Manning, New Riders, McGraw-Hill, Jones & Bartlett, Course Technology, and hundreds more. For more information about Safari Books Online, please visit us online.

How to Contact Us

Please address comments and questions concerning this book to the publisher:

> Make:
> 1160 Battery Street East, Suite 125
> San Francisco, CA 94111
> 877-306-6253 (in the United States or Canada)
> 707-639-1355 (international or local)

Make: unites, inspires, informs, and entertains a growing community of resourceful people who undertake amazing projects in their backyards, basements, and garages. Make: celebrates your right to tweak, hack, and bend any technology to your will. The Make: audience continues to be a growing culture and community that believes in bettering ourselves, our environment, our educational system—our entire world. This is much more than an audience; it's a worldwide movement that Make: is leading—we call it the Maker Movement.

For more information about Make:, visit us online:

> Make: magazine: *http://makezine.com/magazine*
> Maker Faire: *http://makerfaire.com*
> Makezine.com: *http://makezine.com*
> Maker Shed: *http://makershed.com*

We have a web page for this book, where we list errata, examples, and any additional information. You can access this page at *http://shop.oreilly.com/product/0636920031895.do*.

To comment or ask technical questions about this book, send email to *bookquestions@oreilly.com*.

The Paper Inventions Supply Cabinet

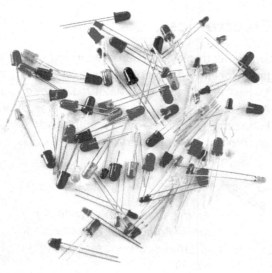

Figure O-1

Most of the materials you need to do the projects in this book can be found in your local crafts supply or hardware store. In fact, you probably own many of them already. The following lists include items that are good to have on hand at home, as well as suggestions for groups and in the classroom. Keep your supply cabinet stocked and you'll always be ready to try some fun and amazing paper inventions!

Paper

For inspiration, your paper stockpile should include as many colors, patterns, and weights as possible. But regular office and children's craft paper is fine for all the projects in this book:

Copy paper:
> for most projects, recycled paper with printing on one side is fine.

Card stock:
> stiff heavy paper with a smooth surface is best.

Construction paper

Origami paper:
> nice but not necessary.

Adding machine paper:
> long thin rolls are handy for some projects but not required.

Recycled newspaper:
> full sized is best.

Recycled magazines:
> glossy home, cooking, and fashion magazines will have lots of colorful photos to use. Ask your local library for old copies.

Art Materials

These are just the basics. Feel free to supplement with your own favorites:

- Markers
- Crayons
- Pencils
- Pens
- Glue sticks
- White glue
- Masking tape
- Clear tape
- Shrinkable plastic sheets (or recycled #6 plastic food containers)

Sewing Supplies

For these items, you may have to visit a specialty shop or order online.

- Embroidery hoop: can be used as a screen for making paper
- Teflon nonstick pressing sheet: used to keep ironing boards clean

Housewares/Groceries/Dollar Store

Kitchen items are easy to find in your local grocery store. Discount and dollar stores can also be a great place to find small electronics you can salvage for parts as well as cheap household items:

- Large plastic basin or aluminum roasting pan
- Sponge
- 2 smooth dish towels (not terry cloth)
- Rice flour
- Potato starch
- Plastic wrap
- Wax paper
- 2 frying pan splatter screens: can be used as screen for papermaking instead of embroidery hoop.

- LEDs: cut from strings of lights or from mini flashlights

Hardware

Most local stores will carry these.

- Aluminum foil tape (near the regular duct tape)
- Copper foil tape (used for gardening); can order from one of the following electronics retailers
- Pliers: for bending wires on LEDs
- Window film: optional for the Paper Generator project

Electronics

Small electronics parts are becoming harder to find locally. Try online retailers such as Makershed.com (*http://www.makershed.com*), Adafruit.com (*http://adafruit.com*), SparkFun.com (*http://sparkfun.com*), and Jameco.com (*http://jameco.com*).

 Note

At the time of this writing, the fate of RadioShack as a source for electronic components was up in the air.

LEDs
Red LEDs are the most versatile, but you can buy a mixed bag of LEDs and test out different kinds.

3 V coin batteries (CR2032)
For a large group, you can buy these in bulk from *cheap-batteries.com*.

Circuit Stickers
These lights and electronic sensors are specifically designed to be used with paper and inspired many of the projects in this

book. Buy them through Makershed or Chibitronics.com (*http://chibitronics.com*).

Rectifier

A component that turns AC current to DC. For the Paper Generator project, a part such as Jameco part number 103026 (*http://www.jameco.com/webapp/wcs/stores/servlet/Product_10001_10001_103026_-1*) will work. (This is strictly optional—the project will work without it.)

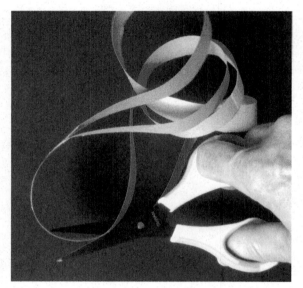

Figure 0-2

Extras You May Find Helpful

Small, sharp, pointy scissors

While you need to be careful, a small pair of really good scissors will make cutting out many of the patterns much easier. For the projects in this book, they can take the place of an art knife.

Ruler

Helpful for measuring and for drawing straight lines. If you choose to use an art knife, it is also useful as a guide.

Paper or foam plates

They can help you keep all your materials and small pieces of paper in one place while you work.

Large ziploc bags

For storing small pieces of paper or finished objects.

Programmable cutter

Many paper projects will go much faster if you own one of these machines. A Silhouette Portrait (*silhouetteamerica.com*) was used to test out the paper machine projects.

Notebook

If you're making changes and trying new things, writing down the details in an "inventor's notebook" will help you remember what worked and what didn't.

Camera

Take pictures while you work to record and share your cool creations!

Paper Science

The invention of paper changed the world! Learn about the science of papermaking, where paper gets its special properties from, and how to use chemistry and physics to make paper models that move.

The Invention of Paper

Figure 1-1 *Making recycled paper by hand produces beautiful results.*

Paper is everywhere, and you see it everyday. So it might be hard to think of it as an invention. But the first people to make paper kept their discovery a secret from much of the world for hundreds of years. They knew they had a valuable new material on their hands.

Before paper, if people wanted to write or draw something and carry it around with them, it wasn't always easy. Soft slabs of clay or wax were popular since they could be erased and reused. Richer folks used silk fabric made from the cocoons of silkworms or parchment made from animal skins. In Egypt, they used the flattened stalks of a plant called papyrus, from which we get the word "paper." But all of these materials were expensive, hard to make, and difficult to use, so they were reserved for only the most important documents.

Then about 2,000 years ago, a Chinese official named Ts'ai Lun came up with a way to make lightweight sheets for writing and drawing that was much quicker and much less expensive. Ts'ai Lun took the bark of a mulberry tree, mixed it with old rags and netting, and mashed it up into sheets of paper.

Paper was so much easier to make, and so inexpensive, that it could even be used for beautiful things like paper lanterns and frivolous things like kites! More importantly, with plenty of paper around, the Chinese began to use carved wooden stamps to print multiple copies of books, newspapers, and even money.

The Chinese managed to keep the formula for papermaking a secret from other parts of the world for a thousand years. And when papermaking finally reached Europe, it took another 500 years before paper replaced fancy parchment. It wasn't until the invention of the mechanical printing press—centuries after the Chinese developed printing—that people in Europe accepted the idea that paper was the wave of the future.

The Science of Papermaking

If you've ever heard of paper referred to as "dead trees," you know what most paper today is made of: wood! Before Europeans reached Central America, the Aztecs were making paper from the inner bark of fig trees. In addition to mulberry bark, the first papermakers in Asia used bamboo, rice straw, and seaweed. But the woody part of trees has only been used to make paper for around the last 150 years. In fact, you can make paper from almost any kind of plant. Up until the mid-1800s, the most common raw materials were old rags of cotton and linen—fabrics that also come from plants (American currency, from the $1 to the $100 bill, is still made from a mixture of cotton and linen). When it became hard to find enough rags, inventors looked for a way to make wood soft enough to use. By the 1860s, a process using special chemicals and machinery allowed paper factories to replace rags with wood to help meet the growing demand for paper.

Whatever it's made out of, the steps for making paper are the same. First, the plant material is cut up, soaked, and mashed to a pulp. The paper pulp is combined with water to create a thin mixture called *slurry*. The slurry is strained through a screen, leaving a thin, even layer of pulp on top. This layer is pressed flat to get rid of excess water and pack the wet pulp together. When it dries, you have paper! If the paper will be used for writing or printing, the paper is also sized. *Sizing* involves adding starch or glue to make paper semi-waterproof, so it won't soak up ink like a sponge.

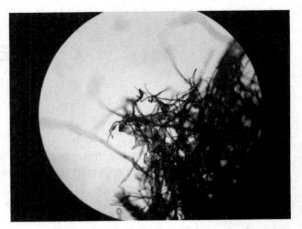

Figure 1-2 *The fibers that make up a piece of paper, seen through a microscope.*

But it's what's inside plants that makes it possible to turn them into paper. Every plant is made up of tiny cells. These cells are surrounded by a tough material called *cellulose*. The cellulose forms long thin fibers that are so small you can barely see them without a magnifying glass. And the fibers are made up of even smaller strings called *fibrils* that can only be seen with a powerful microscope. When plant material is chopped up and soaked, the individual fibers are released. The loose fibers float around in the watery slurry and connect with other fibers. On the microscopic level, wherever the fibers and their little fibrils touch one another, an electrical pull called *van der Waals* forces holds them together.

The more places the fibers touch, the greater the van der Waals bonds between them, and the stronger the paper. But it's easy to break the bonds apart again—just wet the piece of paper! That's why you can take old paper and use it to create new paper. Try it yourself by making your own recycled paper.

Project: Make Your Own Super Easy, Super Quick Recycled Paper

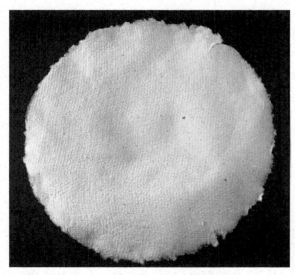

Figure 1-3 *The interesting texture in this handmade paper comes from the cheesecloth used to strain it.*

Scrap paper goes from junk to art when you turn it into handmade recycled paper. This paper uses the dip method, in which a screen is lowered into a vat filled with slurry and lifted out with the wet layer of paper pulp on top. You won't need to add sizing because the old paper you're using to make new paper already contains it.

Materials

- Scrap copy paper (6 to 8 sheets will make several pieces of paper).

- Blender (preferably the kind that looks like a pitcher sitting on a base).

- 2 screens (these can be splatter guards for frying pans from the dollar store, open-weave cloth stretched on embroidery hoops, or window screening or mosquito netting stapled tightly over the opening of small picture frames).

- A vat large enough to hold a strainer, such as a plastic basin or disposable roasting pan.

- Sponge.

- 2 smooth dish towels (not terry cloth).

- Hot tap water—about 3 or 4 cups for each sheet of paper.

- Hair dryer (optional).

- Iron (optional).

Step 1

Tear the scrap paper into pieces roughly an inch wide and two or three inches long. You can also cut it up in a paper shredder.

Figure 1-4

Step 2

Fill the blender, loosely, with the shredded paper. Pour in enough of the hot water to cover it, then let it soak for about 15 minutes.

Figure 1-5 *Start your blender slowly so it doesn't splatter, as it did here!*

Figure 1-6 *Dip your screen into your pan filled with slurry.*

Step 3

Turn the blender on to a slow speed to start mixing the paper and water. As it gets smoother, increase the speed as the paper gets mushy like runny oatmeal. When done, pour the mixture into the vat. Add the rest of the hot water and mix them together with your hands.

Step 4

Take one of the screens and dip it into the vat. Tilt it under the surface of the slurry until it's flat. Slowly lift the screen out of the vat, still holding it flat. You should have a thin, even layer of pulp on your screen. Allow excess water to drain away for a few seconds. If it's the right thickness, go on to Step 5. If it's too thin, mix up more paper pulp and add it to the vat. If the paper's too thick, scoop out some of the pulp with a kitchen strainer. Then dip the screen again.

Figure 1-7

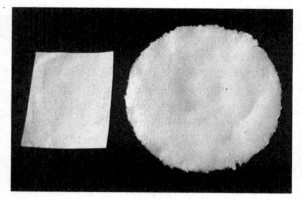

Figure 1-8 *Your paper, cut into shapes*

Step 5

Take the screen and place it on the dish towel. Cover it with the other screen. Press your sponge onto the top screen several times to blot up more water.

Step 6

Remove the top screen and let the layer of wet paper dry on the bottom screen. This may takes hours or a day or more. To speed things up, you can use a hair drier to blow-dry the paper in about 15 minutes. (See Figure 1-1.)

Cut your paper into a rectangle or other shape, or leave its edges soft for a natural effect, as shown in Figure 1-8.

More Papermaking Ideas

To make your homemade paper even more interesting, add some plant material like thin leaves and flower petals, or even colored thread, to the wet paper once you lift it out of the vat. You can even add kitchen scraps like carrot and banana peels.

Or, when you're done making sheets of paper, use a small strainer to scoop up the leftover pulp and turn it into a plantable seed bomb. Squeeze out some of the water and press in a few wildflower or herb seeds. Form the pulp into a ball or disc with your hands, press it into a mold, or place a cookie cutter on a screen and pour or press the paper pulp into it. Blot with a sponge to dry it as quickly as possible, so the seeds don't try to grow before you're ready. When it's dry, toss the seed bomb into any patch of ground you think could use some more greenery!

 Caution!

Don't pour any leftover slurry water down the drain. The paper pulp can clog the pipes. Strain out the solid matter and throw it away separately.

Edible Paper? Why Not?

Paper is commonly made from plant material. And there's no reason why those plants can't be fruits and vegetables! Those photos you sometimes see on birthday cakes are actually printed on wafer paper made from vegetable starch. In China, paper made from rice is used to create edible candy wrappers. And in Vietnam, a different kind of rice paper (actually more like an extremely thin pancake) is a favorite breakfast treat and used to wrap spring rolls and other delicacies.

You can buy edible paper in cake decorating centers and Asian food markets. Try using it to create place cards for fancy dinners, takeout boxes to hold sweets—or for writing secret messages. You can even try making some of the curling, weaving, and building projects later in this book with edible paper. But if you want to keep it edible, make sure your hands are clean as you work, and don't use any non-edible materials in your projects! Here's a quick and easy recipe for making Vietnamese-style rice paper.

Project: Make Edible Rice Paper

Although this rice paper is meant as food, you can also cut it into sheets and use it to write edible messages!

Figure 1-9 *This edible rice paper is colored and flavored with orange juice. The writing was done with an edible-ink marker.*

Materials

- 1 tablespoon rice flour
- 1 tablespoon potato starch
- 1 1/2 tablespoon cold water
- Pinch of salt (optional)
- Small mixing bowl
- Whisk or fork
- Spoon or spatula
- Microwavable plate
- Plastic wrap
- Microwave oven
- Wax paper
- Cooling rack (optional)

Step 1

Whisk the rice flour, potato starch, salt, and cold water together in the bowl for a few seconds. The mixture should be about the same thickness as white glue.

Figure 1-10

Figure 1-11

Step 2

Stretch a piece of plastic wrap across the top of the plate until it is tight like a drum. Pour the mixture onto the plastic wrap. Tilt the plate to spread the mixture around. Try to make a circle at least 7 inches across. The bigger the circle, the thinner your paper.

Step 3

Place the plate in the microwave and cook the mixture on high for 45 seconds. The paper puffs up as the water steams.

Step 4

Put a piece of wax paper on a cooling rack. Use oven mitts to carefully remove the plate, which may be hot. Turn the plate

upside down and put it on the wax paper. Loosen the plastic wrap and remove the plate, leaving the cooked rice paper on the wax paper.

Step 5

Carefully peel the plastic wrap away from the rice paper. Let your edible paper cool.

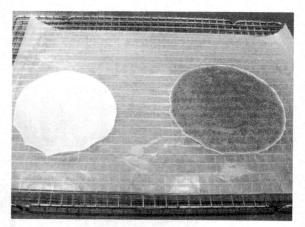

Figure 1-12

Step 6

Your rice paper will curl up around the edges as it dries. Cutting it into a square (using food shears or a knife) helps it stay flat. You can store it for a day or two in a zip-top bag.

Step 7

Variations: for more color and flavor, add a little vanilla, cinnamon, or other spices. You can also replace some or all of the water with something tastier, such as orange juice, maple syrup, or coconut milk. You can also try adding a little mashed banana, strawberry, or blueberry. You may need to adjust the amount of the other ingredients to get the right thickness.

Writing on Your Edible Paper

To write notes on your edible paper, you need edible ink or paint. You can buy edible-ink markers. Or try making your own by boiling down grape or cranberry juice until it is thick. Dip a craft stick or drinking straw in your homebrew ink to write a message. You can also try painting on your edible paper with edible paint made from melted chocolate. Use dark chocolate, milk chocolate, and white chocolate to create different tones.

Self-Folding Paper Models

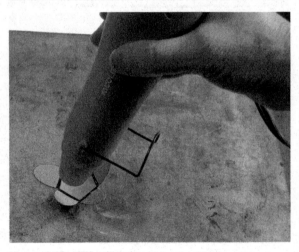

Figure 1-13 *With a little heat you can make a paper flower fold up its petals, all by itself.*

A flat piece of paper that can fold up by itself may seem like magic, but it's really science. In laboratories at Harvard and MIT, scientists are adding "smart materials" like heat-activated shrinkable plastic to paper to create self-folding models. They're even building robots that assemble themselves and walk away!

Shrinkable plastic—the same stuff that's used in crafts materials like Shrinky Dinks—is made from a chemical called polystyrene. The polystyrene is heated until it becomes soft and rubbery. Then it is stretched out into thin sheets and quickly cooled to lock it into shape. When you reheat these stretched-out sheets to just the right temperature, you soften up the plastic enough to allow it snap back to its original size. The result is a smaller, thicker piece of plastic.

To make self-folding paper models using shrinkable plastic, scientists sandwich the plastic between two layers of paper. Then they use laser cutters to cut openings in the paper wherever they want their model to bend. When they heat up the model for a few seconds, only the plastic that's covered by the paper stays cool. Where the paper has been cut away, the plastic can get hot enough to shrink. And as the exposed plastic areas shrink, they pull the sides of the paper model into position.

How to Turn Two Dimensions into Three

It takes *geometry*—the branch of math that deals with shapes—to figure out how to turn a model you can hold in your hand into an outline on a flat piece of paper. A *two-dimensional*, or 2-D, shape has a length and a width. A *three-dimensional*, or 3-D, shape also has height.

In geometry, when you create a 2-D pattern that can be folded up into a closed 3-D shape, you have created a *net*. A net includes all the sides of the object, but it also has to connect them so it can be folded up, which can be tricky. The simplest kind of net is a tetrahedron, which looks like a pyramid made up of four identical triangles. It only has three joints where the paper needs to bend. Next is a cube, which consists of six identical squares and five bendable joints. A three-dimensional shape with flat sides and straight edges like a tetrahedron or a cube is called a *polyhedron*.

Mountain and Valley Folds

Figure 1-14 *Mountain fold on the left, valley fold on the right. Simple!*

Origami is the art of making models from folded paper. It has been studied by scientists, mathematicians, and engineers to help them design things that can be collapsed and reassembled easily. Origami uses the term "mountain fold" to mean folding the paper away from you, so the two sides point down. A "valley fold" creases the paper toward you, so the two sides point up. You'll learn more about origami in Chapter 4!

The other difficult part of creating a self-folding model is making sure the joints all bend in the right direction. To do this, you have to get the inside of the fold to shrink more than the outside and pull the sides of the model inward. One way to control the direction of the bend is by heating the model more on one side than the other. Another is to cut a bigger opening in the paper on the inside of the fold.

In the laboratory, that's easy to do. Scientists use programmable laser cutters to remove exactly the right amount of paper, without touching the plastic layer within. Even if you don't happen to have a laser cutter, you can get the same effect by cutting the paper first along every edge and gluing each piece on separately. It takes a little more time to make a self-folding paper model by hand, but the result is still amazing!

Project: Make Self-Folding Paper Models

Figure 1-15 *These paper models fold up by themselves when you heat them up.*

Use the following templates to make some simple self-folding models. Then get creative and design some of your own!

Materials:

- Cardstock or construction paper
- Shrinkable polystyrene plastic sheets (sold as Shrinky Dinks in art stores; you can also use recycled clear plastic containers marked with recycling symbol #6)
- Computer printer or copier (optional)
- Glue stick
- Scrap paper
- Black permanent markers
- Heat gun (available in craft stores) or toaster oven
- Long, narrow piece of thick cardboard (solid, not corrugated) to use like a spatula in the toaster oven (optional)

Caution!

Paper is flammable, so watch it closely. Heat up your paper model for as short a time as possible. If the paper starts to turn brown or smoke, remove it from the heat immediately!

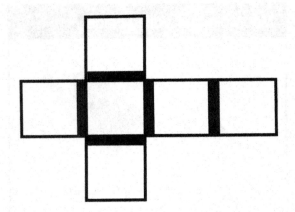

Figure 1-17 *Cube template. For both top and bottom paper layers, cut out the white squares as shown. Cut around the entire shape for the plastic inner layer, coloring in the spaces as shown. Each side of the finished cube is roughly one inch long.*

Figure 1-16 *Tetrahedron template. For the top and bottom paper layers, cut out the smaller white triangles as shown. For the plastic inner layer, cut around the entire upper triangle, coloring in the spaces as shown. Each side of the finished triangle is two inches long.*

Find larger versions of the templates in Appendix A.

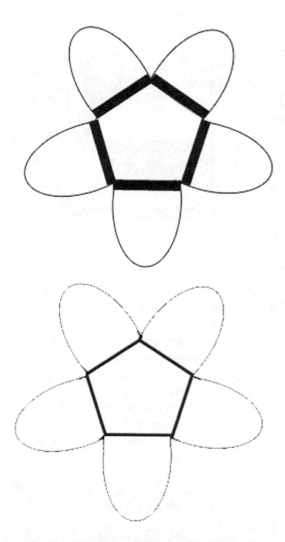

Figure 1-18 *Flower template. For the top and bottom paper layers, cut out the white pentagon center and petal shapes as shown. For the plastic inner layer, cut around the entire upper shape, coloring in the spaces as shown. Each side of the finished pentagon is one inch long.*

Step 1

Choose a shape you'd like to build. For each shape, there's a template for the inside paper layer, the outside paper layer, and the plastic layer in the middle. Copy each template by hand or with a printer. (You can find Shrinky Dinks sheets made for printers,

but other kinds also work. Just be careful not to rub the printer ink off by accident.)

Step 2

Cut around the image you copied onto the plastic. You don't need to cut along the exact lines at this point. Do the same with the paper images. Cut out each side and apply glue:

Figure 1-19

Step 3

Take the outside paper net and separate all the sides by cutting along each edge. Take one piece and place it on the scrap paper with the printed side facing down. Use the glue stick to coat the entire back of the piece with glue. Take the plastic and place it so the printed side is facing up. Attach the glue paper side over its position on the plastic. Do the same with all the other sides.

Step 4

Turn the plastic over. Cut out all the sides of the inside paper image. Cut away the thick black lines. Glue the sides just like before, being sure that the paper pieces don't cover the lines on the plastic.

Step 5

Take the black marker and color over the thick lines where they are exposed. Black absorbs heat, so the black edges should start to shrink before the rest of the plastic.

Step 6

If you have a heat gun, place the paper model with the inside facing up on a heat-proof surface. Turn the heat gun on and blow the hot air on the paper model for a few seconds. Aim for the black lines where you want the model to bend. You should see the sides begin to fold up. Stop as soon as the edges meet. If you wait too long, the sides will start to shrink too.

Figure 1-21

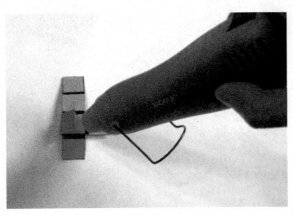

Figure 1-20

Step 7

If you are using a toaster oven, pre-heat it to around 275-300 degrees Fahrenheit. (This is cooler than for regular shrinkable plastic projects.) Place the paper model, inner side facing up, on one end of the thick piece of cardboard. You will hold the cardboard "spatula" while the model is heating up, so make sure it's long enough to keep from burning your fingers.

Step 8

Open the toaster oven door and leave it open. Slide the cardboard holding the paper model onto the rack inside while keeping a grip on the other end. Within a few seconds the sides should start to fold up. If it stops before closing up all the way, pull the model out to make sure the paper isn't starting to burn. If it looks OK, put the model back in the oven for another couple of seconds until the model has closed up as much as it is going to:

Figure 1-22

More Self-Folding Paper Ideas

In theory, any paper model that uses valley folds can be made with this self-folding technique. Just remember that the more sides—and folds—you add to a model, the harder it is for the model to fold itself up.

It is possible to use both valley and mountain folds in one self-folding paper model by cutting larger and smaller openings in the paper, depending on which direction you want that edge to fold. You may have to turn the model over halfway while heating it to make sure both sides of the paper shrink up correctly. (Let the model cool down in between to avoid burning the paper.)

You can also experiment with controlling how far the joint bends by changing the width of the opening in the paper. The amount of bend is known as the *angle*, and it can be measured with a math tool called a *protractor*. See if you can figure out how wide to make the opening to get a bend of 45 degrees or 30 degrees by measuring the finished product with a protractor. (You can find one where you buy school supplies.)

Take notes as you try different versions. That information may help you build more complicated self-folding models!

Paper Tech | 2

Add electricity to create powered paper projects for a whole new level of awesome.

Paper Circuits

Figure 2-1 *You can make circuits that light up with just special tape, an LED, and a battery.*

Building paper circuits is a great way to learn about electricity. They're simple to make, and they can be used to create all kinds of interesting designs! This chapter will show you how to create fun light-up art projects using tiny light bulbs known as LEDs. They're the little lights that tell you if an electrical device is switched on, and they have many other uses as well. But to get started, you first need to understand something about electricity. And it all starts with atoms.

Atoms and Charged Particles

Everything in the universe is made up of extremely tiny building blocks called *atoms*. At the center of an atom is a nucleus made up of smaller particles called protons and neutrons. And surrounding the nucleus is a cloud of even tinier particles called electrons. Protons and electrons carry charges that are equal but opposite, with protons being positive (+) and electrons being negative (-). (Neutrons are neutral and have no charge.) Particles with the same charge repel each other and push each other away. Opposite charges attract each other.

To picture how it works, think about what happens when you put two magnets together. The negative end of one magnetic will pull on the positive end of the other magnet. But if you point the negative ends toward each other, they will push each other away. In their resting state, atoms contain an equal number of protons and electrons, so the positive and negative charges cancel each other out. That means an atom normally has no charge overall.

But electrons can move between one atom and another. Sometimes atoms take on extra electrons and become negatively charged. Atoms can also lose electrons and become positively charged. Charged atoms are called ions. When charged particles or ions move, they produce electricity. (Technically, what they produce is electromagnetic energy, because moving electric charges always produce magnetic forces and moving magnets always produce electric forces.)

Electricity and Circuits

Some materials are better than others at letting charged particles or ions move. They are called conductive. They include metal, salt water, and your skin! A *circuit* is a path of conductive material that allows electrical charge to flow. By creating a circuit, you can direct the electricity to the components you want to get power. But a circuit must be closed to run. That means it it is connected in a loop or a circle (that's where we get the name circuit). Current can also flow to the ground, but you usually don't want that to happen!

A circuit also needs a source of energy, such as a battery, a generator, or a solar cell to get the charged particles or ions moving. The movement of charged particles in a circuit is its *current*. Current is measured in amperes, or amps. The amount of energy available in a circuit is its *voltage*. It is measured in volts. The electrons in a circuit flow in one direction, from the negative end, or terminal, of your power source (-) toward the positive end (+).

To be useful, a circuit needs materials that do not conduct electricity very well, called insulators, to keep the charge from traveling outside the circuit. Paper, rubber, fabric, plastic, and air are good insulators. The insulating ability of a material is its *resistance*. Resistance is measured in ohms.

You can also control the voltage and current in a circuit by raising the resistance, either by using less conductive material or adding a component called a resistor. For example, graphite (like in a pencil) is a poor conductor, so its resistance is high. You can figure out the correct resistance to use if you know the voltage and the current, since the voltage equals the current times the resistance. (If you don't know them already, you can measure them using a device called a multimeter.) This formula, usually written as V=IR, is known as *Ohm's Law*.

The part of the circuit that does the work is called the *load*. For paper projects, these are usually low-power components like lights, speakers, or hobby motors. Components have a positive lead (also called the anode) and a negative lead (also called the cathode). Some components, like motors, will run backward if you connect the positive lead to the negative terminal of your power source. Some components may be damaged if they're connected "backward." And others, like LEDs, won't run at all.

LED (Figure 2-2) stands for light-emitting diode. A *diode* is a special type of electrical component that only lets current flow in one direction. When connecting an LED in a circuit, be sure to point the positive lead toward the positive battery terminal and the negative lead toward the negative terminal. On many LEDs, the positive lead, called the anode, is longer, and the negative lead, called the cathode, is shorter. You can think of it as the "plus" lead had something *added* to it, and the "minus" lead had something *subtracted* from it. Also, the side of the LED bulb is almost always slightly flattened on the cathode side.

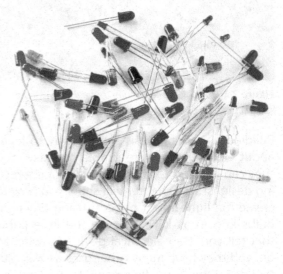

Figure 2-2 *LEDs come in many colors, shapes, and sizes.*

LEDs require a certain minimum voltage to work, and that voltage is different for different colors. LEDs will not light at all if the voltage is too low. If the current is low, the LED will still light, but it will be dimmer. Too much current can damage an LED, but that should not be a problem with the paper circuits in this book.

☀ Caution!

A circuit without a load is called a short circuit. Because pretty much all the energy in the system gets sent back to the power source, it can overheat very quickly. Be careful to avoid creating a short circuit!

How to Create a Paper Circuit

The following are some examples of materials you can use to build a circuit.

Figure 2-3 *Aluminum foil and copper foil tape*

Conductive material

Some conductive materials that work with paper circuits include:

- Peel-and-stick copper tape (available from electronics retailers and hardware stores; comes in 1/4 inch and 1/8 inch widths; look for tape with conductive glue on the back)

- Peel-and-stick aluminum tape (like duct tape, but metallic; can be cut into thin strips)

- Conductive ink like Circuit Scribe (electroninks.com) (*http://www.electro ninks.com/*)

- Conductive paint like Bare Conductive (bareconductive.com) (*http:// www.bareconductive.com/*)

- Graphite (like the lead in very soft pencils)

Insulating material

The paper your circuit is built on can serve as the insulation. If needed, you can also use non-conductive tape (such as electrical tape).

Power source

Most paper circuits can be powered with just a 3 volt coin battery. A good size to use is CR2032. But you can also use two AA or AAA batteries in a battery holder, or a solar panel that generates about 3 volts. If your circuit uses a poor conductor like pencil lines, you may be able to use a 9 V battery by touching it *briefly* to the circuit. (See the following caution box!) And check out the Paper Generator project later in this chapter to see how to power a circuit yourself!

Load

Ordinary LEDs work great for the projects in this chapter. You can buy a grab bag of different types from electronics retailers like those in the following list. Sometimes you will find a few strange shapes in your collection. If you're not sure which end is negative and which is positive, try touching the leads to a coin battery. If it doesn't light up, turn it around. Then mark the positive side for future reference!

Where to Get Materials

- Hardware stores
- Art supply stores
- Makershed.com (*http://www.make rshed.com/*)
- Adafruit.com (*http://adafruit.com*)
- SparkFun.com (*http://sparkfun.com*)
- Jameco.com (*http://jameco.com*)
- RadioShack stores

Project: Build a Basic Paper Circuit

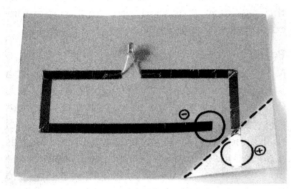

Figure 2-4 *A simple circuit*

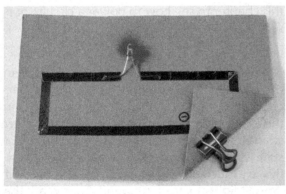

Figure 2-5 *To turn the circuit on, insert a battery where shown, fold the corner over it, and use a binder clip to hold it in place.*

Before you get any further, try making a simple circuit of your own.

Materials:

- Paper
- Copper or aluminum foil tape
- CR2032 coin battery
- LED

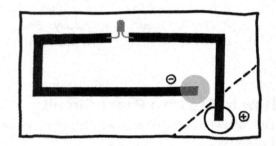

Figure 2-6 *Simple circuit template*

Step 1

Copy the template or draw your own version on a piece of paper. Keep it small so you can use one piece of tape for each section (and avoid having to connect pieces if possible). You will need to fold over your circuit so that one end touches the top of the battery, so put it near the corner of your paper.

Step 2

Now cover the lines that connect the LED to the battery with the metallic tape. When you get to a corner, gently fold the tape into a bend. If you must connect two pieces of tape, fold over the end of the new piece and place it overlapping the first piece. This will make sure that the metal sides of the tape are touching, in case the glue is not very conductive. Don't forget to leave a gap for your LED!

Step 3

Gently bend the legs of the LED as shown, being careful not to snap them off. Place the LED in the gap so that the legs overlap the tape on either side. Use extra pieces of copper tape to hold them down.

Step 4

To turn your circuit on, place the battery where marked and fold over the corner of the paper so that the tape touches the battery.

Troubleshooting Tips

If the LED doesn't light up, try flipping the battery over. If that doesn't work, check to see that there's a good connection between the tape and the LED. Also check any places where two different pieces of tape are connected, and look for tears anywhere along the circuit.

Caution!

Electrical components can be damaged or burst if they receive too much voltage or current. A component called a resistor is often used to avoid going over that limit. The simple paper circuits in this book don't need resistors because they use low power and are only meant to be turned on for a short amount of time.

To keep your paper circuit safe, use a battery or solar panel that produces more than 3 volts only when creating a circuit out of high-resistance materials like graphite. Never try to plug your circuit into the building's electrical supply. You could

cause a fire or give yourself a dangerous shock!

More About Circuits: Adding Switches and Multiple LEDs

The more you learn about circuits and electronics, the more interesting your inventions can become. Here is information about ways to make your designs bigger and more interactive.

Switches

The corner flap on the basic circuit you just built is really just a very simple on-off switch. A switch is nothing more than a gap in the circuit that opens and closes like a drawbridge. When the switch is opened, no current can flow through. When it is closed, the current can flow. In circuits, switches can also redirect current from one path to another.

But switches can be used for more than just turning on lights. In fact, they're at the heart of every piece of electronics. Computers convert all the instructions and information they receive into yes-or-no questions, and represent these two states using millions of microscopic on-off switches.

Series and parallel circuits

There are two main ways to connect multiple switches or components in a circuit. In a *series circuit* (Figure 2-7), the current flows through one component after another. It's like a train on a single track. It has to pass through every station, one after the other, before it returns to its starting point. In a *parallel circuit* (Figure 2-8), the current flows through each component at the same time. It's like a train track that splits so that multiple trains can stop at all the stations simultaneously, then meet up again at the main depot.

Tip

Enlarge the templates so the grey circle is 3/4 inch across, big enough for a CR2032 battery. For larger templates, see Appendix B.

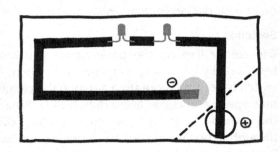

Figure 2-7 *Two LEDs in series*

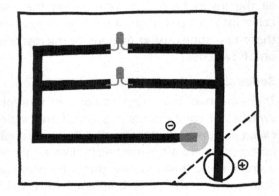

Figure 2-8 *Two LEDs in parallel*

Every time you add another component to a series circuit, the current remains the same but the voltage is split between them. In a parallel circuit, the current is divided among each component but the voltage remains the same. So to light two LEDs in series in your paper circuit, you would need two 3 V batteries. That's why you're usually better off putting multiple LEDs in parallel.

Project: Make a Light-Up Paper Cat

This design takes the basic circuit and adds a second LED in parallel and a switch you press to turn it on. A switch that only stays on while you are pressing it closed is called a *momentary switch* (as shown in Figure 2-9).

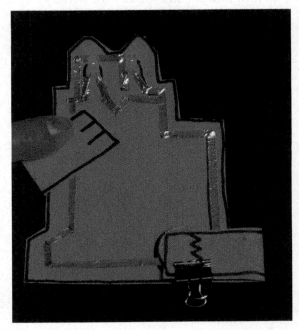

Figure 2-9 *The paw on this Light-Up Paper Cat is a momentary switch. Press it, and the light comes on!*

Materials:

- Heavy paper, like construction paper or cardstock
- Permanent marker
- Copper or aluminum foil tape
- CR2032 coin battery
- LED
- Small binder clip

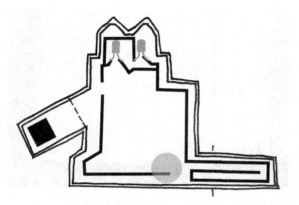

Figure 2-10 *Light-Up Cat template*

Step 1

Copy or draw the design shown in the template onto the paper so the cat is about 4.5 inches high and the battery fits in the marking. Cut it out, leaving a tiny margin of paper around the border. Also draw the border and inner lines on the back of the cat's paw and tail. (See Figure 2-9.)

Step 2

Apply the metallic tape over the circuit lines as shown, leaving gaps for the LEDs and for the switch. Make sure the lines don't touch or cross. If you need to connect two pieces, make sure to fold over the end of second piece and overlap the first. Build up two or three layers of tape to make a pad for the cat's paw. Fold the paw up as shown to make sure the pad closes the switch.

Step 3

Take one of the LEDs and lay it flat where marked. The negative side points to the left. Gently bend the leads downward so they touch the paper, being careful not to snap them off. Use wire cutters or scissors to trim the leads to fit. Use metallic tape to hold them in place. Repeat with the other LED.

Step 4

Place the battery positive side up where marked. Gently bend the cat's tail around so the tip covers the battery. Don't make a

sharp crease. Use the binder clip to hold the tail in place. Make sure the binder clip does not touch the battery!

Step 5

Press the cat's paw down so the pad closes the switch. The LEDs should light up. If not, see the troubleshooting tips in the first project.

Paper Sensors and Special Effects

Sensors are like switches that only turn on under certain conditions. The following are some paper sensors you can build into your circuit.

Simple pressure sensor

Figure 2-11 *The pressure sensor turns the light on when you press it.*

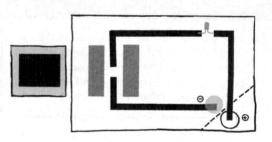

Figure 2-12 *Pressure sensor template*

This sensor works just like the cat's paw switch in the Light-Up Cat circuit. You can use either a folding flap or a separate piece of cardstock. Keep the pieces from touching accidentally by making spacers from peel-and-stick craft foam or wads of masking tape. Make sure the top piece is stiff enough to pop back up when released.

Blinking and flickering effects

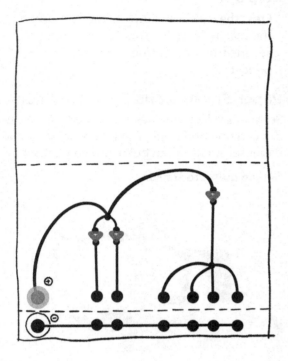

Figure 2-13 *Blinking pressure strip template. Insert the battery and fold up along the bottom dotted line.*

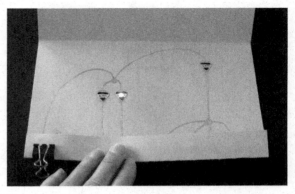

Figure 2-14 *When you press on the pressure strip, the LEDs light up. This circuit is drawn with a Circuit Scribe conductive ink gel pen, but you can substitute metallic tape instead.*

Figure 2-15 *The card is closed and the lights are hidden inside. When you run your finger along the bottom, the owl's eyes blink, and the star in the upper right flickers.*

You can use pressure sensors to make your lights go on and off in different patterns. Fold up the edge of a piece of cardstock to make a narrow flap. Inside put a row of pressure sensors, connected to multiple LEDs. Close the flap and run your finger along the pressure strip to open and close the circuits.

Chibitronics

This card was inspired by a design created by Jie Qi using Chibitronics Circuit Stickers, stick-on LEDs that are flat and let you fold down the top of the paper to make a greeting card that glows from inside. (*http://chibitronics.com/*)

Magnetic switch

Figure 2-16 *Place a magnet underneath this switch to pull the paper clip down and close the circuit.*

A *reed switch* has two contacts that close when a magnet is near. Make a paper version by taping a paper clip (or other small metallic object that is attracted to magnets) on top of a regular flap or pressure switch. When you hold a magnet underneath, the switch should close.

Project: Make a Paper Tilt Sensor

A tilt sensor uses some kind of movable conductive material, such as liquid mercury or steel balls, to close a switch. When the sensor is tilted, gravity pulls the conductive material down so it bridges a gap in the circuit. This tilt sensor is housed inside a folded paper "pillow box."

Figure 2-17 *The jingle bell closes the circuit inside the Tilt Sensor.*

Materials:

- Copy paper
- Copper or aluminum foil tape
- Clear tape
- CR2032 coin battery
- LED
- Jingle bell (from crafts store), metal button, copper penny, or aluminum foil wadded up into a ball, about 1 inch in diameter

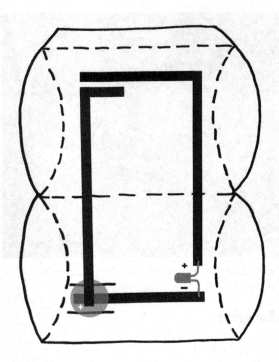

Figure 2-18 *Tilt sensor template*

Step 1

Copy or draw the template onto the piece of paper so that the box is about 6.5 inches long and the battery fits in the marking. Cut along the solid lines. This design uses slits in the paper to hold the battery in place instead of a binder clip. Carefully cut open the slits as shown on the template.

Step 2

Fold up the box along the dotted lines. Follow the curve of the lines on the sides. When you tuck them in, the box should look like a little bed pillow. Unfold.

Step 3

Lay the metallic tape along the circuit line that connects the battery and the LED. Next, fold over about one inch of tape so that it sticks to itself to make a flap that will go on top of the battery. Then run the rest of the piece of tape up to the switch area, bending at the corner. Then lay the third piece of tape as shown. Make sure to leave space between the two pieces of tape that run next to each other at the switch area.

Step 4

Place the LED where shown. It should lie flat against the paper. Make sure the leads overlap the metallic tape on either side. If necessary, bend and/or trim the leads to fit. Then use the metallic tape to attach it securely to the circuit.

Step 5

Slip the battery into the slits you cut, positive (smooth) side up. Fold the flap of metallic tape on top. Secure it with clear tape. Then cover the entire circuit EXCEPT for the "switch" area with more clear tape.

Step 6

Refold the box. Use tape to hold the straight side and one curved side securely closed. Open the other curved end and insert the metallic ball, then close it up again. Test your tilt sensor by turning it so that the ball rolls over the "switch" area. The LED should make the paper box glow!

More Paper Circuit Ideas

You can extend your paper circuit by adding different components, including:

- A small buzzer (from an electronics retailer, or recycled from a dollar-store keychain).
- A speaker (you can find recordable speakers in talking greeting cards).
- A vibrating motor (these can be found in old cell phones or disposable electric toothbrushes).
- Shape memory wire to make your paper invention curl up when the circuit is turned on (see the book *Making Simple Robots* for directions).

There are also kits that can help you build paper circuits, such as:

- Circuit Stickers, pre-glued components that include LEDs, sensors, and special effects stickers with pre-programmed blink patterns. Circuit Stickers were invented by Jie Qi of MIT's Media Lab, whose paper circuits inspired the projects in this chapter! Available from *chibitronics.com* and Maker Shed (*http://makershed.com/collections/circuit-stickers*).

- Circuit Scribe, a kit that lets you draw circuits using a pen with conductive silver ink. Circuit Scribe kits also come with components like LEDs and motors that you connect using a magnetic board. Available from *electroninks.com*.

- Activating Origami, which comes with all materials including motors and LEDs. Available from *teknikio.com* and Maker Shed (*http://makershed.com/products/activating-origami-kit*).

Paper Generators

A paper generator is a wild invention. When you tap or rub it, it creates enough static electricity to light an LED!

Figure 2-19 *You can light your LED paper projects by rubbing the aluminum foil circuit with a charged sheet made of Teflon.*

You've probably felt a little shock of static electricity, also called triboelectricity, when you pulled socks out of the dryer, or scuffed your shoes across a carpet and touched a doorknob (or a friend). You can see the electrical field it creates when you run a comb through your hair in dry weather, causing your hair to stand on end! The friction of the comb gives each strand of hair the same electrical charge; and like magnetic fields, similar electrical charges repel each other. Your hair stands on end because each strand of hair is trying to get away from all the other strands of hair!

The paper generator works by building up a static charge in an electret—a material that can be made to hold a permanent charge, just like a permanent magnet holds onto its magnetic field. In this case, the electret is a sheet of Teflon. You normally think of Teflon (also known as PTFE, or polytetrafluoroethylene) as the non-stick coating on pots and pans. But it also has the ability to hold onto extra electrons, which give it a negative charge.

When you put a sheet of PTFE on top of a circuit and rub or tap it, it creates a field of static electricity that can light up an LED. The simplest version of a paper generator can only make the LED flash on and off. That's because your rubbing motion makes the current flow back and forth. That is called alternating current, or AC. An LED will only light up steadily when the current is going in one direction. That's called direct current, or DC. However, the voltage produced by a paper generator is high enough to light up multiple LEDs in series.

The idea for paper generators came from scientists at Disney Research. They used it to make test versions of self-lighting greeting cards, books, and game boards. You can see a video of their prototypes at *disneyresearch.com/project/paper-generators*.

Project: Make a Paper Generator

Figure 2-20 *Create a paper generator that lights up when you tap it.*

This project includes directions for two different homemade versions of paper generators. The LEDs will only glow dimly, so you may need to test your paper generator in a darkened room. Once you've got the hang of it, come up with your own ideas for human-powered inventions!

Materials:

Figure 2-21 *Nonstick PTFE (Teflon) is the only special material required for a paper generator, but pieces of silver-coated polyester film (such as Mylar) work slightly better than regular aluminum foil.*

- PTFE (Teflon) sheet (sold at fabric stores as "nonstick pressing sheets" for ironing for under $10—enough to make 12-24 generators)
- Newspaper
- Aluminum foil tape or copper foil tape
- Silver-coated polyester film or PET (sold at hardware stores as mirrored window film for about $20—a 15 x 3 foot roll is enough for 45 generators) (optional)
- Cardstock or stiff paper
- Adhesive dots (like Glue Dots) and/or masking tape
- LED--preferably white or blue with a clear bulb (try the LED from a dollar-store garden light)
- For tapping generator: peel-and-stick craft foam (or wads of masking tape)

Version 1: Rubbing Generator

This paper generator consists of aluminum tape, an LED, and a small Teflon sheet:

Step 1

Place two full-width strips of aluminum foil tape (or wide strips of aluminum foil, or mirrored window film, held on with adhesive dots or masking tape) on a stiff piece of paper, about a quarter inch apart. Bend the leads of an LED at right angles and tape one lead onto each strip of foil, so the LED is pointing straight up.

Step 2

Cut a piece of PTFE sheet about 3 1/2 by 4 1/2 inches. Crumple up a sheet of newspaper and rub it on the nonstick sheet 20 or 30 times. This charges the sheet by letting it steal electrons from the newspaper!

Step 3

Place the generator on a table in a darkened room. Put one hand on one strip of foil so your body closes the circuit. With your other hand, rub the nonstick sheet on the other strip of foil. You should see the LED flicker rapidly. It may be very dim at first, but the longer you rub, the brighter the LED appears.

Step 4

Another way to light the LED is to place the PTFE sheet on the foil and tap it. You should see a flash of light every time you strike the foil.

Version 2: Tapping Generator

This tapping generator consists of a PTFE sheet sandwiched between two pieces of mirrored window film that conduct the static electricity generated to the LED:

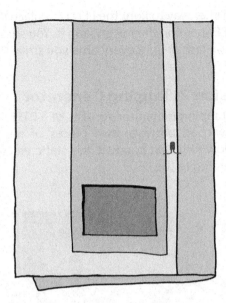

Figure 2-22 *Tapping Paper Generator template*

Step 1

Copy or draw the design shown on the template (Figure 2-22) onto the cardstock. Carefully cut out the window, about 2 1/2" high x 3" wide.

Step 2

Turn the paper over and use masking tape to attach a piece of PTFE sheet about 3 1/2" x 4 1/2" over the window:

Charge the sheet by rubbing it with crumpled-up newspaper 20 or 30 times.

Step 3

Cut a piece of window film or aluminum foil about 8 inches by 11 inches. Place it over the PTFE sheet so that the top and bottom line up with the cardstock, and one edge hangs off the edge of the cardstock by about 1 1/2". Use adhesive dots and/or masking tape to attach it to the paper:

Step 4

Turn the paper over. Fold the window film over the edge and attach it with adhesive dots:

Step 5

Cut two strips of craft foam about 4 inches by .5 inch. Place one on either side of the window:

Step 6

Cut a second piece of window film about 4 inches by 10.5 inches. Attach the bottom of the film or foil to the craft foam strips and the top of the cardstock base with adhesive dots. The foam should keep the top layer of film from touching the Teflon sheet, unless you press down on it.

Step 7

Cut a 4 by 4 inch square of cardstock to make a tapping pad. Use adhesive dots to attach it to the window film right over the window:

Attach the sides to the cardstock with masking tape. Attach an LED to the strips of window film so that it bridges the gap between the two strips of film or foil. In a darkened room, tap on the square of cardstock. The LED should flash on and off.

More Paper Generator Ideas

Test whether you can improve your paper generator with different designs, different materials, or different components. Possibilities include:

- Silver-coated polyester (PET), sold at home improvement stores as mirrored window film (a 15' x 3' roll costs about $20 and is enough for 45 generators).

- Other metallic-looking materials, like the Mylar used in balloons, wrapping paper, and snack bags.

- LEDs of different colors and shapes (remember to connect them in series, not parallel, if using more than one).

- Mini neon light bulbs (available from Jameco.com, part number 27351 (*http://www.jameco.com/1/1/751-ne-2-wire-terminal-neon-lamp-a1a.html*)).

- Adding a rectifier—a component that turns AC current to DC—in the circuit between the rubbing pads and the LED (such as Jameco part number 103026 (*http://www.jameco.com/webapp/wcs/stores/servlet/Product_10001_10001_103026_-1*)):

Figure 2-23

Paper Engineering

3

Paper is an excellent material for prototyping at low cost with a minimum of tools, but it's also used to make imaginative structures and entertaining machines.

Paper Structures

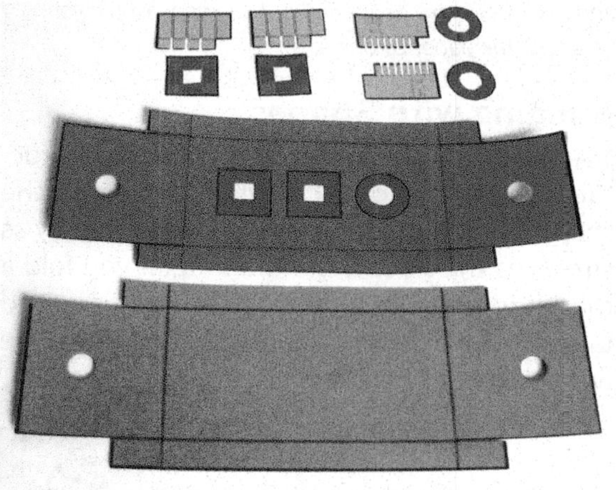

Figure 3-1 *Paper machines are made of cogs, widgets, and do-dads, just like other kinds of machines.*

You probably don't usually think of paper as a building material. After all, a piece of paper sags under its own weight. It certainly can't hold anything else up, like a heavy pile of books!

But when you bend a piece of paper, something interesting happens. All of a sudden, you've got a strong, stable material—strong enough to support a pile of books, or even a person!

In fact, paper can be used to create all kinds of 3-D objects, from pop-up books and toys to furniture and houses. Designers who use paper as a building material are called *paper engineers*. Paper engineers know as much about their ma-

terial as engineers who build bridges know about steel. No matter what they're designing, all engineers have to think about what their material can and can't do, and how to make it work the way it's supposed to. But at the same time, paper engineers try to make their structures good-looking and fun to use. It's a big challenge! Here are some tips to help you get started with paper engineering.

Paper Under Pressure

As you've already learned, paper is made up of tiny fibers. Those fibers are held together tightly by strong microscopic forces. However, that strength only works in one direction. If you take a piece of paper by the ends and try to stretch it until it tears, you'll have to pull really hard. Paper is really good at standing up to stretching forces. Engineers call that stretching force *tension*. But if you lay a pencil across a piece of paper and pick the paper up loosely by both ends, it sags in the middle. That's because gravity is pulling the pencil down, and the paper is too thin to stand up to the force of the pencil pushing on it. That pushing force is called *compression*.

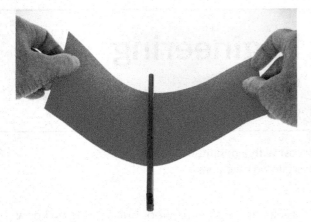

Figure 3-2 *Hold a piece of paper loosely by the ends and lay a pencil across it. The paper will sag.*

Figure 3-3 *Fold the same piece of paper accordion-style and it will hold up a pencil without sagging, even if you only hold the paper by one end!*

How can you help paper stand up to compression forces? One way is to make it thicker. A thick sheet of cardboard can hold up a pencil when paper can't. If you glue enough sheets of paper together, they'll hold up a pencil, too.

Another way is to make use of the strength from other parts of the paper, not just the middle. *Distributing* the weight means spreading that compression force around so it's not just pushing on one area. You can do this by simply folding it!

To see how it works, take the piece of paper from the pencil experiment and fold it up like an accordion. (In other words, fold it back and forth so it zigzags between mountain folds and valley folds.) Now lay the pencil crosswise on top of the folds. The paper won't sag because the weight of the pencil is shared by both sides of each fold. It's like you're gathering up strength from other parts of the paper and packing into one small area. Paper or other material that is folded this way for added strength is called *corrugated*.

Building with Shapes

You might think that the best shape for a structure is a square or rectangle, since that's the shape of most buildings. But a square is only as strong as the sides that hold it up. If you fold a thin strip of paper into a hollow square and tape it, it will stand up on its own. Add a little weight, and it's easy to make it lean to one side.

Figure 3-4 *A pencil on top of an empty paper square makes the square sag. The bottom of the square keeps the sides from sliding apart, so it leans to the side instead.*

When you put weight on a circle, the entire circle helps to hold it up. That's why a rounded arch inside a square opening like a door or window can help the square keep its shape. Try rolling a strip of paper into a circle and fitting it inside the square. See how it makes it harder to lean the square over? The rolled-up strip of paper is under tension: it wants to spring back open. That tension helps balance the compression the square is feeling from the weight on top.

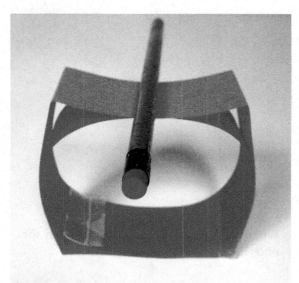

Figure 3-5 *Add a circle inside and it helps keep the square upright.*

The strongest shape of all is a triangle. The sides support the compression of the triangle's weight, and the bottom of the triangle pulls on both the sides. The tension across the bottom balances the compression on the sides, making a triangle very stable. Fold a strip of paper into a triangle, and you can press on it much harder before it begins to bend or lean over.

But no matter what shape your paper structure is, you can make it stronger by folding down the edges. That creates a flap that you can connect with tape or glue to the sides next to it.

The more folds you add, the stronger your paper structure will be. If you roll or fold a sheet of paper up into a thin stick, you can use it to build very strong paper structures. A thin rod, beam, or stick used to build a structure is called a *strut*. The following are some interesting structures you can build using just paper struts and tape or glue!

Project: Make a Tabletop Tetrahedron

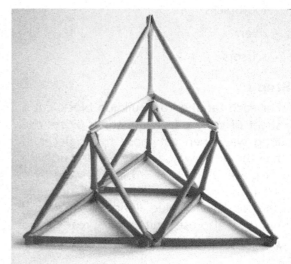

Figure 3-6 *The triangles in this paper tetrahedron help give it strength and stability.*

A tabletop tetrahedron is like a mini pyramid. To build it, you need to construct lots of smaller tetrahedrons and stack them like building blocks. In mathematics, the model you are building is known a *fractal*, a design made up of smaller copies of itself. Each tetrahedron is made of six struts made of rolled paper. To make a two-level pyramid, you need four smaller tetrahedrons, or 24 struts. A three-level tetrahedron takes eight smaller versions, or 48 paper struts. Want to add more levels? You do the math!

Materials:

- 8 1/2 by 11 inch copy paper (one sheet for every one of the smaller tetrahedron "building blocks")

- glue stick (for a more permanent structure, use white glue)

- small adhesive dots (1/8 inch is good)

- bamboo barbecue skewers or round toothpicks

- ruler

- scissors

Step 1

For each tetrahedron building block, cut a sheet of copy paper into three pieces the long way. Then cut those three pieces in half the other way. You should end up with six pieces that are 3 3/4 inches wide by 4 1/4 inch long. Take the glue stick and apply an even layer of glue along one short end (you can also use white glue spread with a toothpick):

Figure 3-7

Lay a bamboo skewer across the opposite end. Start to roll the paper up as tightly and

smoothly as possible, keeping the ends even. When you reach the glued strip, press the paper roll closed for a few seconds with your hands:

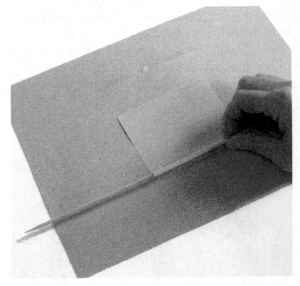

Figure 3-8

Step 2

When you have made six struts, pinch the ends flat and a little bent:

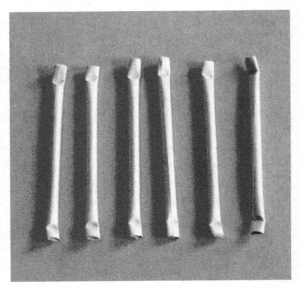

Figure 3-9

Take three sticks and lay them flat in a triangle. Apply adhesive dots to the end of one strut and attach it to the end of the next strut. Do the same with the other corners:

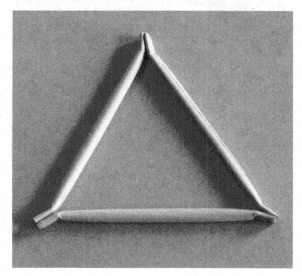

Figure 3-10

Step 3

Take three more struts and attach one end to each corner of the triangle with more adhesive dots (or dabs of white glue). The new struts should be pointing upward toward the center of the triangle. Apply adhesive dots to all three ends and press together. Your first tetrahedron is finished!

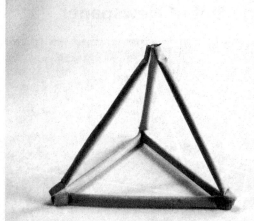

Figure 3-11

Step 4

To keep adding to your tetrahedron fractal, make three more small structures. Place them on your work surface, corners touching, so that their bases form a larger triangle. Use adhesive dots to connect the corners that touch. Then put an adhesive dot on the top of the tetrahedrons on the first level and place the fourth tetrahedron on top. If you wish, continue to add layers the same way.

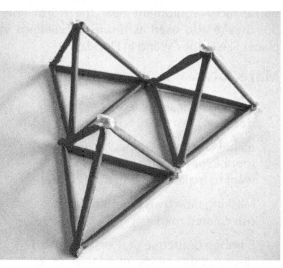

Figure 3-12 *A completed fractal*

Project: Make a Geodesic Dome Out of Newspaper

- Yard stick
- Scissors

Figure 3-13 *This geodesic dome would make a cool place to hold storytime at the library.*

Figure 3-14 *Use the corner of a table to help you roll your newspaper in a straight line, starting with the corner.*

Geodesic domes don't just look cool. They're also way stronger than regular building shapes. Like tetrahedrons, they're made completely of triangles. But they also have the strength of rounded arches. And they don't need any internal walls or supports to hold them up, so they use a minimum of materials. Geodesic domes are so strong and compact, they're used to house radar equipment near the North Pole. But they're also used as futuristic buildings in places like Disney World in Florida.

Materials

- Enough floor space to assemble your dome, at least 10-12 feet across
- 65 full-size sheets of newspaper (double or triple that number if you want to make your structure sturdier)
- Masking tape (two different colors, or use colored markers)
- Bamboo barbeque skewers, 1/8 inch diameter dowels, or round toothpicks (can be left in or reused)

Step 1

Lay a sheet of newspaper on the table. (Use two or three sheets if you want to make your struts extra strong.) Place a toothpick or skewer at one corner, and tuck the corner of the newspaper under it. Then use it to help you roll the sheet up as tightly as you can to form a strut.

Tip: To help you roll at the correct angle, place the corner of the newspaper on the corner of your work surface, and roll straight ahead.

As you roll, gently slide your hands apart to keep the ends nice and tight. When you reach the other corner of the newspaper, wrap it tightly around the middle with a piece of masking tape. Repeat until you have 65 struts.

Step 2

Next, use the scissors to trim about an inch off one end of each strut. Then use a yardstick to measure the struts to the proper length and trim off the other end. You will need:

• 35 long struts that are 28 inches (71 cm) each

• 30 short struts that are 26 inches (66 cm) each

Mark the long struts and the short struts with different color tape or markers so you can tell the two sizes apart.

Step 3

Now, begin to build your dome. Take three long struts and tape them together at the ends to form a triangle. Make four more triangles, for a total of five. These are the long triangles.

Step 4

Make five more triangles the same way, but use one long strut for the base and two short struts for the sides. These are the short triangles.

Step 5

The base of the dome is a decagon with 10 sides. To make it, you will lay down all the triangles you just created so that their bases form a rough circle. Start by laying down one long triangle. Now lay a short triangle next to it, so that one end of the base (the long strut) is touching a corner of the other triangle. Continue alternating long and short triangles around the rough circle, tops pointing in toward the center, until they are all touching.

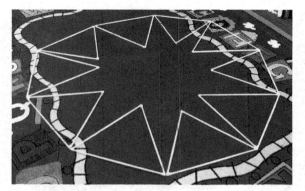

Figure 3-15 *Connect all the triangles around the decagon with the tape.*

Here's a diagram of that first row:

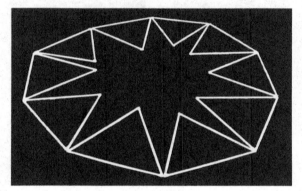

Figure 3-16

Step 6

Take a short strut and use it to connect the top corners of one triangle to the top corner of the one next to it. It helps to do this with a partner: one person to hold the strut, and one person to tape. Go around the dome and connect all the triangle tops the same way. The first level of the dome should now be standing up and leaning a bit toward the center.

Figure 3-17

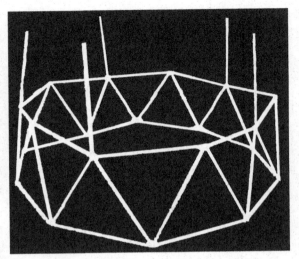

Figure 3-19

Then use two long struts to connect the top of the loose stick to the top of the triangle to the right and the left. Repeat all around the dome:

Figure 3-18 *Use as much tape as needed to hold your joints together.*

Step 7

For the next level, tape one end of a short strut to the top of every short triangle:

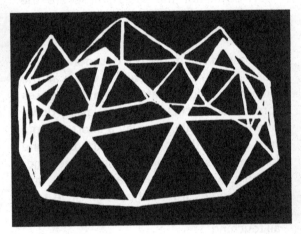

Figure 3-20

At this point, it's a good idea to inspect your dome for any broken or loose connections. Wherever corners of triangles meet, loop some tape through the openings from one triangle to another until every opening is secured.

Step 8

To make the last level, take five long struts and lay them end to end to form a pentagon. Tape them together. Then tape one short strut to each corner and let them flop into the middle. Take all the loose ends and connect them with more tape. Then fit the pentagon into the opening at the top of your dome. Secure everything with plenty of tape:

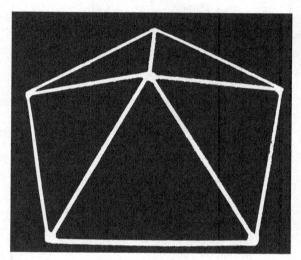

Figure 3-21

Step 9

If you like, you can cover your dome with flat sheets of newspaper to create a playhouse or shelter.

Buckminster Fuller

The geodesic dome was invented by an architect named Buckminster Fuller. To learn more about his designs, check out the tensegrity robot project in the book *Making Simple Robots*, also from Maker Media.

Paper Machines

Figure 3-22 *A prototype for an automaton uses masking tape instead of glue so parts can be easily changed around.*

Animated paper models are really paper machines. Another name for a moving paper model is an *automaton* (plural: automata). The word is also used for early wind-up robots built for entertainment. In Japan, moving mechanical paper toys are called *karakuri*. The first karakuri were made in the 1600s and are still popular today.

Most animated paper models are hand-powered. To make them move, you have to turn a shaft. A *shaft* is simply a rod that connects the "engine" of the machine to the parts that move. Almost any kind of moving part or mechanism can be created out of paper. These parts can make the paper model move up and down, rock back and forth, or spin around. Combining different mechanisms lets you create very interesting movement, and really make paper models come alive.

One of the simplest kinds of mechanisms is the cam. A *cam* is a disc that turns around the shaft and pushes or pulls on other moving parts of the machine. Cams can be shaped like circles, squares, triangles, or eggs. A snail cam looks something like a spiral snail shell. Cams can also have multiple points, like a star, or bumpy edges. Each shape produces a different kind of motion.

Key to the Paper Model Templates in This Book

- Cut along solid black lines.
- Cut out solid black circles, squares, etc.
- Dotted lines are mountain folds; dashed lines are valley folds (see the "Mountain and Valley Folds" box in Chapter 1).
- Shaded (gray) areas show where to glue on tabs or guide bases.
- Tabs on the shaft, rods, and guides should be cut apart but not folded. After the piece is rolled or folded into shape, glued shut, and inserted into its base, the tabs get folded out like the petals on a daisy and glued to the base.
- The side of a square shaft or rod marked "Diagonal Flap" gets folded first. This flap should be folded only partway, so it forms a triangle with the square walls of the shaft or rod. This helps support the shaft or rod from the inside.
- Small circles or squares with crossed lines in the middle of a piece are holes with flaps that get glued to the shaft. Cut along the crossed lines, then fold back the little triangular tabs that are formed.

Tips for Designing and Building Paper Models

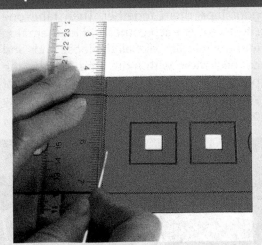

Figure 3-23 *Score the lines on the paper before folding using a dried ballpoint pen or other rounded pointy object, such as this letter opener.*

- On most paper model patterns that you buy or print out, solid lines are cut lines, and dotted lines are fold lines. Mountain folds and valley folds are shown with different patterns of dots. (See the box in Chapter 1).
- Sharp, small-bladed craft scissors are excellent for making paper models. Some paper engineers prefer X-acto knives. If you use a knife, always use a metal-edged ruler to keep straight lines straight. A cutting mat or piece of thick cardboard is recommended, even with scissors, to protect your work surface.
- After you cut openings, use a pencil, craft stick, or toothpick to smooth the edges so they don't rub as the rods move through them. You can carefully trim any bits of paper sticking out with your scissors, but don't make the hole too big.
- *Scoring* is a paper-building technique that makes it easier to fold the paper in the right place. To score the fold lines, press down along them using a tool with a rounded tip. Dried-up ballpoint pens are good for scoring, as are plastic knitting needles. Anything with a

rounded point that is easy to "draw" with will work. Use a ruler for straight lines, and go slowly and carefully. Be sure not to cut or tear the paper as you score it.

- Glue sticks are handy for covering large areas with glue. For small areas, scrape off a tiny amount with a flat toothpick and spread it where needed. White glue is used by most experts, but it should be spread as thinly as possible. Squeeze out a blob on a piece of scrap paper to work from, and spread it with a toothpick where needed. Always give your model time to dry between steps and before testing it with moving parts! You can use paper clips or binder clips to hold pieces closed while they dry.

- When trying out new designs, masking tape is a quick and easy way to put a model together. Whenever you need to make a paper model stronger and more stable, add a diagonal strut to create a triangle.

Project: Make a Paper Cam Sampler

Figure 3-24 *The rods on this paper machine go up and down or spin around when you turn the handle. You can use them to make your own automated scene.*

This Cam Sampler demonstrates the kinds of motion produced by three different mechanisms: an egg cam, a snail cam, and a friction drive. You can mix and match these shapes or modify them any way you like to design your own moving paper model.

Materials

- Cardstock
- Sharp craft or fabric scissors with short blades
- Cutting mat or thick cardboard
- Ruler
- Glue stick (or white glue, spread thinly with a toothpick)
- Scrap paper to put under pieces when spreading glue across them
- Craft stick and round pencil (to slide inside square and round tubes as they are glued)
- Paper clips and/or binder clips (to hold glued flaps closed as they dry)
- Masking tape

You can find larger versions of these templates in Appendix C.

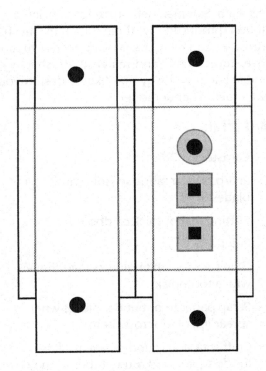

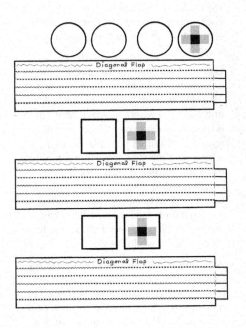

Figure 3-25 *When folded into shape, the cam sampler base is 5 inches long, 3 inches wide, and 3 inches high. The side flaps are 1/2 inch deep.*

Figure 3-26 *When folded into shape, the rods are 6 inches long, 1/4 inch wide, and 1/4 inch high. The taps are 1/4 inch long.*

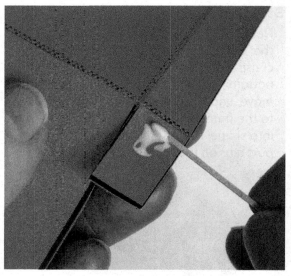

Figure 3-28

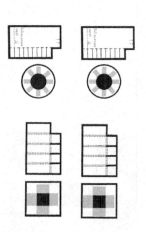

Figure 3-29

Figure 3-27 *When folded into shape, the square guides are a little less than 3/8 inches on a side. The round guides are about 1/2 inch across.*

Step 1

Copy the templates by hand or with a copier. Each will fit on an 8 1/2 by 11 inch piece of paper.

Step 2

Cut out and assemble the frame that will hold all the machinery and models. It comes in two parts, upper and lower. Build each one separately. You will connect them at the very end. Fold along the dotted lines as shown. Spread a thin layer of glue on the side tabs. Hold the glued parts together for a few seconds to make sure they are secure. Wipe away any excess glue and let dry.

Step 3

Next, build the square guides that go on the upper part of the frame. Each guide consists of a tube with support tabs on the bottom to hold the rods in place as they move, and a flat base to connect the guide to the frame. Before folding the guide tube into shape, make sure to cut apart the support tabs on the bottom.

where the tabs should be attached). Poke the tube into the hole in the base facing the glued side. Slide it down until the tabs are touching the glue. Spread a little more glue on top of the tabs, and then press the whole thing over the matching opening in the frame.

Figure 3-30

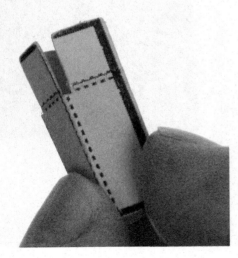

Step 4

Once the guide tube is assembled, fold the support tabs outward. To attach the guide to its flat base, spread glue on the top of the base (the side with the markings that show

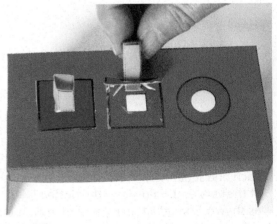

Figure 3-31

Step 5

The friction cam gets two round guides. One will go on top of the frame, and the other will go underneath. Build each round guide as you did the square guides by rolling it up starting at the end opposite the glue flap. You can use a round pencil or pen to roll the paper around and to give you something to press against when gluing it closed. The round guide fits into its base the same as the square guides. When the guides are assembled and dry, glue one to the top of the frame next to the square guides. Glue the other to the underside of the frame, lining it up with the round guide on the top.

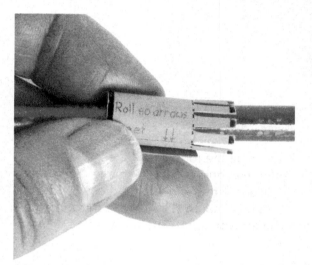

Figure 3-32

Figure 3-33

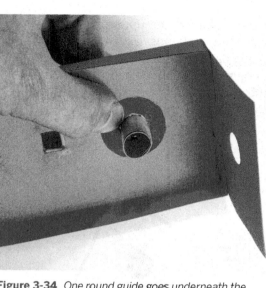

Figure 3-34 *One round guide goes underneath the frame to hold the rod straight as it turns.*

Step 6

The rods make the animated parts of the machine move. They have flat feet, also called cam followers, that get pushed by the cams as they rotate. To build a rod, cut the support tabs apart and fold the rod into shape, starting with the diagonal flap. Fold it up as smoothly and tightly as possible so it can move inside the guide without getting stuck:

Figure 3-35

Spread glue on the upper part of the foot with the markings, as you did with the guide base. Poke the top of the rod into the opening in the upper part of the foot until the tabs are touching the glued surface of the foot:

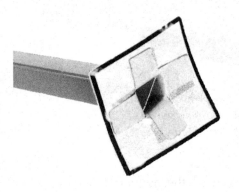

Figure 3-36

Spread a little more glue over the tabs and press the lower part of the foot over them, sandwiching the tabs in between.

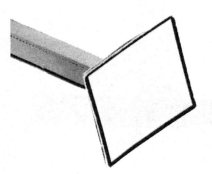

Figure 3-37

Figure 3-38

When all the rods are finished, insert them into the guides on the frame as shown. Make sure they move up and down (or turn around) smoothly. If not, try stretching the guides with a craft stick or pencil until they do.

Figure 3-39

Step 7

Now cut out and assemble the drive shaft. When you turn it, all the machinery moves. The tabs at the end are where you will attach the handle when the entire machine is finished. Fold up and glue the shaft the same way you did with the rods.

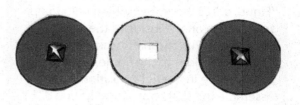

Figure 3-41

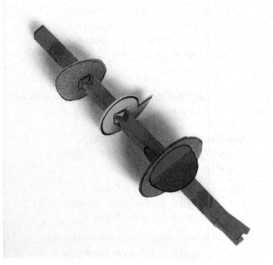

Figure 3-40

Step 8

To assemble the egg and snail cams, you will glue two matching pieces together. First cut open the crossed lines in the center and fold back the triangle-shaped tabs. This creates an opening that will let you attach the cam to the shaft. Spread glue on one side of the cam and glue it to the other. Make sure the openings line up and the triangular tabs are facing outward. The round friction cam is made the same way, but it is four pieces thick. Glue the inner pieces (without the tabs) together first. Then glue one piece with triangular tabs facing out to each side.

Step 9

Fit the cams on the drive shaft, but don't glue them yet. The round friction cam goes at the end with the tabs, the snail cam goes in the middle, and the egg cam goes on the other end. The egg and snail cams should be pointing in opposite directions. Insert the ends of the drive shaft into the holes on either side of the upper frame. Slide the rods into the guides with the feet underneath. Adjust the cams so the egg and snail cams are directly under the square foot of the rod above them. Glue in place. The friction cam should be placed so it is close to the edge of the round foot. Try turning the shaft to make sure the rods move as they should. Glue in place. To help keep the friction drive cam straight on the shaft, take a plain round piece and fold it in the middle. Glue it to the friction cam and the shaft as shown. Rotate the shaft one-quarter of the way around and glue another round support to the cam the same way.

Step 10

Take the shaft out of the frame. Slide the upper frame over the lower frame, with the walls of the upper frame on the outside (Figure 3-42). When everything is lined up, spread glue on the inside of the bottom frame walls. Put the shaft back through the holes in the sides of the frame. Make sure the cams are in the right positions. Assemble the shaft stops just like you did the cams. Slide them on the ends of the shaft until they are close to but not touching the walls of the frame. Glue them to the shaft as you did the cams. The turn handle goes on the end of the drive shaft the same way the foot went on the rods.

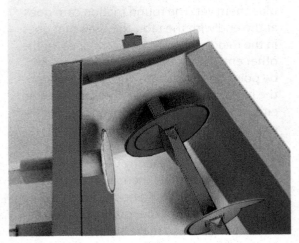

Figure 3-42 *Fit the upper part of the frame over the lower part.*

Got everything working? Now comes the fun part! Use your Cam Sampler to add life to a scene you build with paper models. You can put separate models on each cam, or use more than one cam on one model to create more complicated movements. The next project is an example of using all three cams on one model.

Automata Troubleshooting

If your Cam Sampler isn't working just the way you want, you may need to make some adjustments. Be sure that:

- Your rods are thin enough to move around in the guides without rubbing.

- Your cams are straight up and down. If they are leaning over or starting to bend, glue on some extra layers of cardstock or add a support to help hold it in place.

Getting the friction drive cam to work smoothly can be very tricky. Here are some things to try:

- Make sure the large cam wheel is touching the underside of the smaller spinning disk.

- Make sure the spinning disk is completely flat. Both wheels should meet like the lines in the letter *T*.

- If the friction (rubbing) between the two wheels is not enough to make it turn every time, think about ways to make them stickier, such as putting rougher paper, masking tape, or peel-and-stick craft foam on one or both disks. You could even cut out teeth around the edges so they fit together like gears.

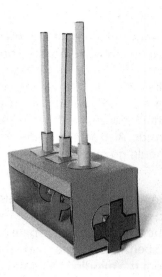

Figure 3-43 *The paper machine is ready for some automated characters.*

Project: Build a Paper Space Rover Automaton

Figure 3-44 *A test model for the Space Rover*

This model Space Rover is designed to fit on the Cam Sampler base. The two up-and-down cams make the rover "body" appear to bounce across an extraterrestrial landscape, while the friction drive cam makes the camera "head" spin slowly so it can take a good look around. For this design, it doesn't matter if your friction drive doesn't spin perfectly: the random starts and stops make it seem like it's scanning the horizon for alien life.

You can find a larger version of this template in Appendix D.

More Paper Automaton Ideas

- Make a wind-powered paper automaton powered by a paper bellows or a straw you can blow through.

- Add a DC motor to turn the shaft, or LEDs to make parts light up. The electronics kits from littleBits (littlebits.cc) have parts you can use, including dimmers to make the motors run faster and slower.

- Cut up clean, recycled paper milk or juice cartons to create a waterwheel that can turn your shaft.

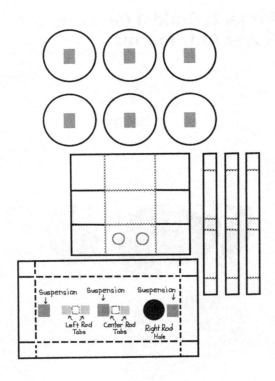

Figure 3-45 *When folded into shape, the rover body is 4 3/8 inches long and 2 inches wide. The side flaps are 1/2 inch deep. The head is 1 1/2 inches long, 1 inch wide, and 1 inch high. The suspensions are 4 1/4 inches long (unfolded) and 5/8 inches wide. The wheels are about 1 1/8 inches across.*

Materials

- Cardstock templates
- Scissors
- White glue or glue sticks
- Markers

Step 1

Build the frame of the Cam Sampler, including the rod guides, following the preceding directions. Then assemble the shaft and cams the same way as you did with the Cam Sampler. Don't put the shaft in the frame yet.

Step 2

Copy or draw the templates for the Space Rover. Cut out and assemble the body of the rover by folding down the sides and gluing the tabs to hold them in place. The markings that show where to glue the wheel attachments and the rods should go on the inside. Also assemble the head. The "eyes" should go on the outside.

Step 3

The rover has three pairs of wheels that are attached to the underside of the rover body by bouncy rocker arm suspensions. Fold each suspension so the two ends hang down from the middle at an angle, like flapping wings on a bird. Glue the wheels to the ends of the arms where marked. Then glue the flat middle section of each suspension to the underside of the rover body where marked. Make sure not to cover the opening for the rotating camera "neck."

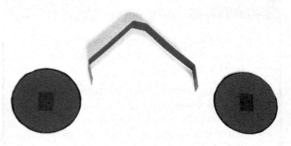

Figure 3-46 *The rover wheels attach to a suspension piece.*

Step 4

Assemble the rods for the Space Rover, but don't attach the "feet" yet. Two of the rods are shorter and have only two tabs at the top. Fold the tabs on those rods back and glue them to the underside of the body where marked. Note that one of the tabs for the center rod overlaps the center suspension.

Figure 3-47 *The underside of the rover body, which shows how the wheel suspensions and the cam rods are attached.*

 Note

The rods that go with the Space Rover are different than those on the Cam Sampler. If you've already built the original version but want to use it as the base for this model, cut off the top of the rods for the egg cam and snail cam so each rod so is 3 1/4 inches long. Then for all three rods, snip down along the corners to make tabs that are 1/4 inch deep. Fold down the tabs and cut or pinch off the diagonal support piece inside. For the egg and snail cam rods, remove two of the tabs opposite each other.

Step 5

Insert the center rod through the center guide on the base frame, and insert the end rod through the end guide for the other up-and-down cam. Take the longer friction drive rod and glue a square support piece to the top. When dry, center the support piece on the underside of the head and glue it in place. Then insert the rod through the round hole in the rover body and through the round guide on the base frame.

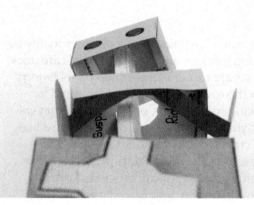

Figure 3-48

Step 6

When all three rods have been inserted through the guides on the top of the frame, glue the "feet" on the ends of all three rods. Finally, insert the shaft, line up the cams and the cam follower feet, and make sure the friction drive discs are touching. Spin the shaft to test that everything's working, then glue the end pieces and the handle on. Your Space Rover is finished!

Figure 3-49

Step 7

You can give your rover more personality by using markers and extra pieces of cardstock to create camera "eyes," solar panel "wings" on the rover's back, bendable "arms" for taking soil samples, or an antenna that will let your rover phone home. Or if you'd like to create an environment for your Space Rover, you can try to design a larger base frame and wrap a cardstock backdrop around it. Then you can cut out or draw a rocky landscape that makes it look like your rover is rolling past the hills of Mars.

Figure 3-50 *The head with its camera eyes slowly scans the horizons for signs of alien life.*

More on Animated Paper Models

There's a lot more to learn about building paper automata, and many more kinds of mechanisms you can build. Here are some good resources to get you started.

Paper automata designers and their websites

- Rob Ives (*robives.com*)
- Walter Ruffler (*walterruffler.de*)
- Keith Newstead (*keithnewsteadautomata.com*)
- Robert Addams (*mechanical-toys.com*)

Books

- *Paper Models That Move: 14 Ingenious Automata, and More* by Walter Ruffler (Dover Publications, 2010)
- *Karakuri: How to Make Mechanical Paper Models That Move* by Keisuke Saka (St. Martin's Press, 2010)
- *Paper Engineering & Pop-ups for Dummies* by Rob Ives (Wiley Publishing, Inc., 2009)
- *How to Design and Make Simple Automata* by Robert Addams (Craft Education, 2006)

3-D Paper Art

4.

Paper bends, twists, and curls into all sorts of useful shapes. Create wearables and household decorations using colorful new or recycled paper.

Figure 4-1 *Paper makes a great art material in itself.*

But that's not the only way paper can serve as an art medium in its own right.

Making paper by hand (which you tried in Chapter 1) is considered an art form because of the unique results you can get by varying the thickness, adding color, and embedding decorative elements like leaves and flowers. Three-dimensional paper sculptures are sometimes made to look like real objects or living things, like the models you created for your paper machine in Chapter 3, but other paper sculptures consist of pure abstract shapes. You saw in Chapter 2 how using paper's ability to let light shine through is another way to make it part of the picture.

In fact, it's very easy to combine the artistic use of paper with science, technology, engineering, and math. Here are some ways to create original and fun works of art and crafts using two traditional methods—curling and weaving—that are very modern at the same time.

Paper as an Art Material

You can make art by drawing or painting on paper. And you can make art out of paper itself. The most well-known example is origami, which you will learn more about in Chapter 5.

Curling Paper into Art

Project: Learn to Make Quilling Shapes

Figure 4-3 *Some of the many shapes you can create with quilling.*

Quilling is the art of creating 2-D and 3-D designs out of thin paper spirals and curls. The strange name comes from quills, stiff narrow tubes taken from feathers, originally used to roll up the paper strips. In quilling designs, the curled shapes fit together like puzzle pieces. A design made up of many different shapes fitted together is called a *mosaic*. Some experts believe quilling dates back to medieval times, when religious women recycled the gold-tipped edges of pages from holy books. We know that in the 1700s and 1800s, wealthy young ladies learned quilling along with other arts like needlework, music, and watercolor painting. Magazines and pattern books were filled with instructions for quilling designs that could be used to adorn furniture or pictures. Today, quilling is used in scrapbooking, greeting cards, jewelry, and household decorations.

Figure 4-2 *Paper can be twirled into all kinds of interesting shapes.*

You can buy quilling tools and quilling paper in all colors. But special tools aren't really needed. All it takes is everyday craft materials and some imagination. And the technique can be mastered in minutes.

You only need three things for quilling: paper strips, a tool for curling the paper, and glue to hold it in place. Professional quillers use medium-weight scrapbook paper with a very smooth finish. You can make your own quilling strips by cutting up pieces of construction paper the long way using scissors or an office paper cutter. You can also run the sheets of paper through a straight-cut shredder. A good width to start is about 1/4 inch, but you can cut different widths by hand to make the coils in your sculpture stand out at different heights.

Materials

- Paper strips (store bought or home made) in different colors

- Scissors

- Wax paper

- Glue

- Cocktail straw (cut a 1/4-inch slot in one end) or quilling tool

- Flat toothpicks

Step 1

Before you begin, set up your work area. Lay out a square of wax paper like a place-mat to protect your work surface and keep the glued quilling from sticking to the table:

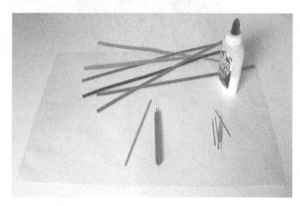

Step 2

Squirt a small mound of white glue onto a corner of the wax paper.

Step 3

To make your own quilling tool, cut a slit in the edge of a cocktail straw. The slit should be as long as your paper stips are wide.

Step 4

To start quilling, choose a strip of paper. Slip one narrow end of the paper into the slit on the straw's tip to hold it in place. Begin to turn the straw, wrapping the paper strip around it evenly and tightly, until the strip of paper looks like a dollhouse-sized roll of toilet paper:

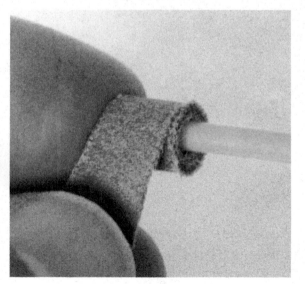

Step 5

Carefully slide the roll off the tool and let it spring open to get the size coil you wish:

Step 6

For a closed coil, dab the tip of the toothpick into the glue and use it to spread a tiny amount on the loose end:

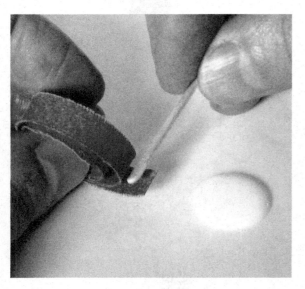

Step 7

Press the glued end where you want it to stick and hold it in place with your fingers for a few seconds until the glue begins to dry.

Allow the coil a few more seconds to dry. Then pinch and mold the closed coil with your fingers into whatever shape you like. Here are some ideas:

Curved shapes

Pinch closed round coils at one or two points to create teardrops, pointy leaves, crescent moons, and waves.

Straight shapes

Squeeze a closed round coil on all sides at the same time into a straight-sided shape like a triangle, square, or rectangle. Leave one side round to create a semi-circle.

More shapes

Leave the coil loose and roll each end separately. With open coils, you can create *S* shapes, flower stems, and sinewy vine effects. Pinch a sharp valley crease in the middle, roll both ends toward the center, and glue the two half-coils together to create a heart.

To create a solid disk, glue the coil closed without letting it open after rolling. You can eliminate the hole in the center if you roll the coil up without a tool. Wetting the end will help you get started.

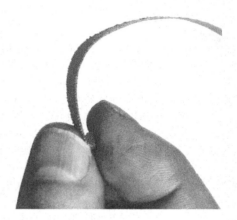

Cones

To create a cone, take a disk and push in the center with your finger, a rounded point like the cap of a pen, or a toothpick. The other side will poke out. If you make the cone high enough, you can gently bend it like the horn of a rhinoceros.

What can you do with your quilling shapes? You can simply "doodle" by connecting them with more dabs of glue into random patterns. Or draw the outline of a shape on a sheet of paper and fill it in with quilled shapes as you go. Finish it off by using more dabs of of glue to connect your quilled design to the piece of paper. You can also go 3-D with your quilling and form your curls into a freestanding bowl or animal figure.

Note

There are no rules in quilling! You can cut your strips shorter, or glue two or more together to make one longer piece. Combine colors in one curl, or double or triple the strips for something thicker. For a design made of separate loops of increasing length (see the blue row on Figure 4-3), you can wind a strip of paper around the teeth of a pocket comb, catching a loop on the next tooth down each time. Slip the loops off carefully and glue in place. You can even cut loops open for or snip the edge for a fringe-like effect. Anything goes!

Project: Make a Quilling Sun Spiral Design

Figure 4-4 *This sun-inspired quilling design can be hung by a window or worn as a necklace.*

Make a simple design inspired by the surface of the sun, with swirling areas of gas reaching out from the surface.

Materials

- 6 strips of paper: 2 yellow, 2 orange, and 2 red
- 1 large yellow closed round coil, about 1 inch across
- 4 medium orange closed round coil, about 1/2 inch across
- 11 small closed round coil, about 1/4 inch across (pink, yellow, and light purple are good)
- 3 red cones, about 1/2 inch high
- 2 violet half moons
- 3 red half moons
- 3 violet waves
- 3 red waves
- 1 blue disk with a hole in the center large enough for a string
- Extra shapes as needed
- String for hanging the finished design

Step 1

Take a strip of yellow paper and loop it around into a ring about three inches across. Glue it shut. Loop another strip of yellow around the first, dabbing glue between them to hold them in place. Then add two orange and two red strips the same way:

Step 3

Around the outside of the loop, glue the blue disk to the top of the sun. Then glue on the half moons and waves as shown. Add more if needed to cover the entire loop. Poke some string through the hole to hang your sun in front of a window or around your neck!

Step 2

Starting with the large yellow circle in the center, arrange the other coils within the tricolor ring you just created. Thicker or thinner paper will make the disks larger or smaller, so add more disks or coils as needed. When you like the arrangement, use dabs of glue to connect the coils to each other and to the outer ring. Where you have dabbed glue, squeeze the ring to make sure it touches the coils inside. That may distort the circle a little, making it more egg-shaped than round. But as an artist, you can change the look of your subject to suit your tastes!

Note

You can seal your finished quilling designs for protection from moisture by painting them with watered-down white glue, Mod Podge, or other paper sealant. Test the sealant on a scrap piece of quilling to make sure it works with your paper.

Weaving Paper

Weaving is a process of connecting strands or strips of fiber, fabric, or other material over and under each other. The result can be stiff or flexible, but often it is also stronger than a single layer of the material alone. Weaving also creates interesting patterns and textures that can be very simple or very intricate. In fact, the first "programmable" machine, the Jacquard loom, was invented in 1801 and used to automatically create fabric designs by weaving threads of different colors and sizes.

Paper can also be woven into mats, bowls, and wearables. Sometimes the paper is rolled into a flexible "rope." For other projects, the paper can be folded into flat strips. The projects here use flat strips of paper that meet each other in a "T" shape. Elements of a design (or lines in math) that go side to side are called *horizontal*. Those that go straight up and down are *vertical*. When horizontal and vertical elements or lines meet, they are said to be *perpendicular* to each other. And they are still perpendicular even when they are tilted or diagonal, as long as they meet in a *T*.

Project: Make a Woven Paper Wristband

Figure 4-5 *Origami paper is used here to create a fun wristband from folded strips.*

In the 1960s, kids used this weaving technique to turn old chewing gum wrappers into bracelets. In 2015, a Virginia man named Gary Duschl set the Guinness World Record for the longest gum wrapper chain ever. It was 83,625 feet long and contained 2,000,000 gum wrappers. But you don't have to use candy wrappers. You can make this woven wristband from any kind of paper that folds easily, including origami paper and recycled glossy magazine pages.

Materials

- Several sheets of thin paper
- Ruler
- Scissors

Step 1

Cut out strips 1 1/2 x 4 inches. You will need about 24 to 30, depending on how big around your finished bracelet is. This is one-sided origami paper:

Step 2

Take one strip and fold it in half lengthwise. Unfold. Take each long edge and fold it toward the middle, so that the two edges meet at the middle fold line:

Fold in half again along the original middle fold line:

Fold the strip closed:

Step 3

Do the same thing in the other direction:

Fold the strip in half so that the short edges meet:

Unfold. Take each short edge and fold it toward the middle so that the two edges meet at the middle fold line. Close up along the original fold line. You now have a V-shaped link:

Step 4

Make a second *V*-shaped link, then attach it to the first link. To do this, take the first link by the point of the *V* and hold it so the single-fold side is facing you. Take the second link by the point of the *V*, double-fold side up. Slip the arms of the second link through the openings inside the arms of the first link. Push the second link all the way down to the point at the bottom of the *V*. The second link should be pushed up against the top of the arms of the first link:

Step 5

Make a third link and slide it into the arms of the second link from the bottom up, creating the zigzag shape. Continue adding links until the chain is big enough to make a loop that can slip over the wearer's hand comfortably without falling off:

Step 6

To close the loop, make sure the last link fits into the first link (in other words, if one is zigging up, the other should zag down). Unfold the arms of the last link so they stick straight up. Place the first link inside the arms of the last link so that they join up like the other links:

One at a time, refold each arm, inserting it into the first link so that it looks like the other links. The ends of the last link get inserted into the first link and then tucked inside:

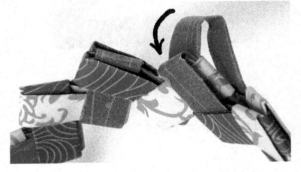

You're finished!

Tips for Weaving with Paper Links

- If you are adding a new link over an old one, slide it down so that arms of the old link fit snugly in the bottom of the V shape of the new link. If you are adding a new link through an old one (as you will do in the next project), slide the tops of the arms of the V between the folds of the arms of the old link.

- If you have trouble joining links because they are too wide, make them narrower by either cutting the strips of paper a bit thinner or overlapping the edges on the first lengthwise fold. Keep in mind that the more snugly the links fit, the better the chain holds together.

- It is easier to insert a new link into an old one from the side that only has a single fold. For this project, insert new links with the double-fold edge up, and you will always have the single-fold side ready for the next link.

Project: Open Weave Paper Link Basket

Figure 4-6 *A basket made from recycled magazine pages.*

Use the same kind of links as the Paper Woven Wristband to make a shiny and colorful basket from recycled glossy magazines. Bags and baskets made from woven snack wrappers have been popular for some time, but this basket's pattern of openings gives it a unique look. The pattern for this basket was adapted from a project on the website D.I.Y. (*http://bit.ly/1MY8qyL*) Done! It's Yours!.

Materials

- Magazine pages, preferably 11 inches long
- Ruler
- Pen
- Scissors
- Mod Podge or white glue and paint brush (optional)

Note

The directions here will make a tall narrow basket. To make the basket wider, add more links to the bottom section. You can use this technique to make a tote by adding a shoulder strap after you finish the top edge.

Step 1

Take one magazine page and cut it into strips that are about five times as long as they are wide. A good-size strip to work with is roughly 2 1/4 inches wide x 11 inches long. (If your magazine is about 9 x 11 inches, you can just cut it into four equal strips.) Fold the strips into *V*-shaped links by following steps 1 through 3 of the Paper Woven Wristband directions. Make about 50 strips to get started, and add more as needed.

Step 2

The design of the basket consists of squares made of four links woven together. To make the first square, hold one link vertically (straight up and down), so the *V* shape is upside down. Hold a second link horizontal-

ly (sideways) with the point of the *V* shape to the right. Slide the arms of the horizontal link over the arms of a second link (see steps 4 and 5 of the Paper Woven Wristband directions):

Take a third strip and slide it over the arms of the first link. It should be next to but in the opposite direction of the second strip:

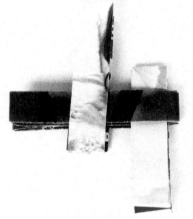

Slide the fourth strip over the second strip so it is next to the third strip but pointing in the opposite direction:

sure the strips go over and under each other in the order shown here:

Then "lock" the square by sliding the arms of the fourth link *through* the arms of the first strip and pulling all the links tight:

Step 3

Start to make the next square by building on the first square. Slide a link over the right arm of the first square, so the V shape is upright.

When you're finished, the links of your square should alternate over and under and each side of the square should have the end of one link sticking out like an arm. Make

Then slide a second link over that link, pointing away from the existing square. Finally, slide a third link over the second and through the first, with the point of the V facing up. Tighten the links until that square is locked in place. Make a third square the same way.

ond the same way. This section of nine squares will be the bottom of the basket.

Step 4

Start a second row of three squares below the first row by following the same steps as before:

Use the arm that points down from the first square as your starting point, and connect the new squares to the row above as you go. Then make a third row below the sec-

Step 5

Now for the tricky part: make four corners that will help the sides turn up. To do that, you will be sliding the arm of four of the squares into the arms of the square next to them, as shown in the diagram:

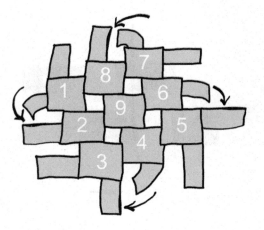

First, find the first square (number 1 on the diagram). Take the arm of the square directly below it (number 2 on the diagram) in the second row.

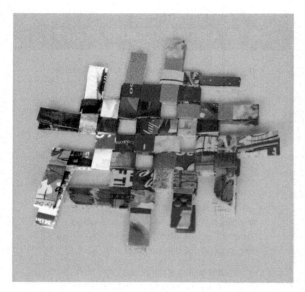

Slide a link over the arm, pointing down. Pull the top row up toward you until it is standing upright. This is the first side. Take the left arm sticking out from the square labeled 1 on the diagram. Bend the arm so that it slides through the left arm of square number 2 (the arm that you added a strip to in Step 5).

Finish the new square with a third link that slides over the bottom link and through the one above (the arm of the first square that you bent around the corner) and tighten to lock it:

When you're done, there should be two sides of the basket standing up, perpendicular to the bottom of the basket and to each other:

Turn the basket around so the next side is on top, and follow the same steps to make the third corner. Do all the corners before filling in the missing parts of the row.

Step 6

Now you have the beginning of four sides standing up from the bottom, with top arms that point toward each other. Go back and build squares on the loose arms, connecting neighboring arms. Continue adding rows until the basket is the height you want.

To continue building up the sides of the basket, take a pair of arms that are pointing toward each other, like the red striped and yellow arms shown here:

Add a new link and start building squares as before:

As you finish each square, you'll be creating more arms to keep building upon:

The finished basket has a round bottom and a pattern of openings that make it interesting looking:

When your basket is as tall as you'd like it to be, finish off the top edge. Take one link with its arms pointing up. Take the arm on the inside of the basket and tuck the end between the two arms (down inside the "wall" of the basket). Bring the end of the outside arm over the top of the basket and over the inside arm. Tuck it

into the loop of paper for that link. A little bit of the end will stick out below the loop. You can add a dab of glue to keep that end in place. Your basket is now ready to hold your favorite possessions!

Paper Math | 5

When you fold, cut, and twist paper into knots, it follows rules that mathematicians are just beginning to discover.

Figure 5-1 *A hexaflexagon.*

Math isn't just about times tables and word problems. One of the things mathematicians do is to find formulas to describe the world around them. And over the years, paper has been a convenient material for investigating and explaining different ideas in math. Even today, when computers can perform calculations millions of times faster than the human brain and create an image of almost anything, mathematicians still use paper models to help them see, touch, and understand the way the world works.

Erik Demaine of MIT is one researcher who has focused on using paper models in math. He's come up with many equations that describe what will happen when you fold, twist, or cut a piece of paper in certain ways. These kinds of formulas are being used in the design of everything from spacecraft to robots. And some mathematicians turn their paper models into works of art. Engineering professor George Hart (father of mathemusician Vi Hart, famous for her math videos on YouTube) makes sculptures out of interesting mathematical shapes. Erik Demaine's father Martin creates curved origami sculptures, and writes about math with his son. But you don't have to be a genius to use paper math models. Even beginners can learn new concepts by creating them. They're fun to build and play with and are a great tool to get you thinking like a mathematician!

Paper Fractals

As mentioned in Chapter 3, a fractal is a pattern that is made up of smaller copies of itself, like the multi-level rolled paper tetrahedron. It's easy to see how the tetrahedron shape repeats itself. But many fractals are incredibly complicated, making it hard to tell what pattern they are forming. Some of these fractals are so detailed that they resemble clouds, leaves, feathers, or other natural objects.

The dragon curve fractal has the shape of a giant fire-breathing lizard, but it's made up of one line that bends back and forth, creating what look like piles of boxes. This fascinating design was invented in 1966 by two NASA scientists, John Heighway and William Harter, who were simply playing around with a folded dollar bill. Heighway got the notion that folding a dollar bill in half over and over would produce a random series of bends without any pattern. When they tried it with a longer piece of paper, they discovered that with enough folds, they could create a curly pattern that looked a little like a Chinese dragon. And each time they folded the paper, they created another version of the same dragon-like shape. In math, repeating the same steps over and over is called *iteration*.

There are limits to the number of times you can fold a piece of paper, but you can program a computer to draw a dragon curve by giving it a mathematical formula that tells it when to make the line turn left and when it should turn right. As the number of iterations gets higher and higher, you will notice that the dragon curve is made up of smaller dragon curves stuck end to end—a classic fractal!

Dragon curves have a number of interesting features:

- They are space-filling curves. In other words, when you add more folds to a dragon curve, the outline of the shape doesn't just get bigger. The bendy line that makes up the shape starts to fill in the empty areas between bends in the previous version, until it almost looks solid.

- If you take four dragon curves and rotate them so each one points in a different direction (up, down, left, and right), they will fit together like puzzle pieces. In math, a pattern made of repeating shapes that fit together is called a *tiling* or *tessellation*.

- If you put two dragon curves back to back so the head of one is touching the tail of the other, they will also fit together.

Even though dragon curves are more of a mathematical oddity than a useful structure,

they have turned up in some unusual places. In the book *Jurassic Park* by Michael Crichton (the source for the hit series of movies about dinosaurs brought back to life), a dragon curve appears on the first page of every chapter, and it gets longer and more complex as the book goes on. Crichton probably used the image to give readers a hint about what was to come: a story where playing around with extinct creatures produced unexpected results that were beautiful and complicated like fractals but scary like dragons at the same time.

The Math Behind Folding Paper

The number of times you can fold a piece of paper in half is surprisingly small. For many years, the limit was believed to be seven or eight, but in 2002 high-school student Britney Gallivan folded a piece of paper in half 12 times. The sample she used was a 4,000-foot-long piece of toilet paper. By the 11th fold, the pile of paper was only a couple feet long and less than a foot high. Gallivan also wrote an equation explaining how to figure out the limit to the number of folds possible. The equation would help you find the minimum length of paper needed to create a certain number of folds, depending on the thickness of the paper. Gallivan's discovery proved that any amount of folds are possible, if the paper is long enough. However, once you reach the limit for a particular length of paper, you need a piece four times as long to add one more fold!

Project: Fold a Paper Dragon Curve Fractal

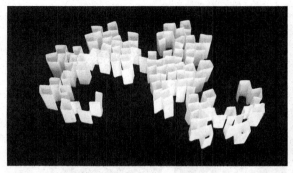

Figure 5-2 *The dragon curve fractal was discovered by scientists who folded some paper over and over.*

To make your own dragon curve, it helps to use paper that is thin but stiff. A roll of long, thin adding machine paper from an office-supply store works well. But you can also cut up regular paper into equal strips and tape them end-to-end. Try not to overlap the ends of the strips as you tape them, to make the strip easier to fold.

Materials

- Thin strip of paper, at least 3 feet long
- Clear tape

Step 1

Making a dragon curve couldn't be easier. Simply fold your long strip of paper in half by taking one end and bringing it over to the other. Make sure the ends are even, then sharpen the crease by pressing your finger along it. Next, fold the paper in half again in the same direction, by bringing the fold over to where the two ends are stacked. Repeat until you have folded the paper in half seven times, or as many times as the thickness of your paper allows.

Step 2

Now unfold the last fold that you made, and put the strip of paper on its side so you can see how it curves. It should form a first-order curve, meaning it has one fold in it:

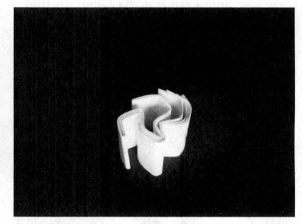

Figure 5-3

Unfold the next fold to make a second-order curve. Sharpen any creases that need it so that every bend in the paper is a right angle. A *right angle* is formed by two lines or surfaces that are perpendicular, like the corners of a square.

Step 3

Keep unfolding the folds, sharpening the creases, and arranging the bends so they form right angles. You should start to see your dragon curve take shape with each iteration. By the time you get up to the fourth-order curve, some of the bends will come together to form a closed box:

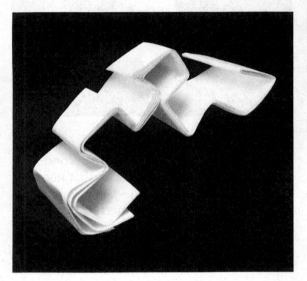

Figure 5-4

By the fifth-order curve, you will get more boxes that are closed off by portions of the paper that turn and come back:

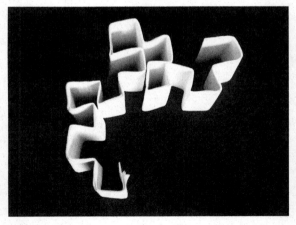

Figure 5-5

If you got as high as seven or eight folds, your dragon curve will start to fill in, making the dragon shape more visible:

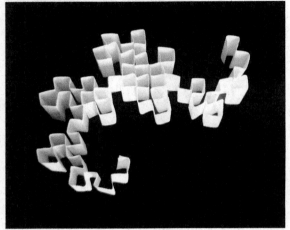

Figure 5-6

Step 4

If you'd like to keep your finished dragon curve, use tape to hold the closed boxes in place. To do this, put a short piece of tape on one side of the box where two different bends meet so it hangs over the edge. Then bring the other bend over and attach it with the rest of the piece of tape. If there is another place where different bends meet, do the same thing there as well. It may help to fold the tape back before you try to bring the bends together.

Figure 5-7

The Amazing, Versatile Box Pleat

Most traditional origami models start with one of a handful of different basic folds. But there's one fold, the box pleat, that can produce any shape possible. The proof of the adaptability of the box pleat was developed by mathematician Erik Demaine. He was also part of the team that used the box-pleat pattern to create self-folding robots using shrinkable plastic sheets covered by paper, like the project in Chapter 1.

Project: Action Origami Robot Worm

Figure 5-8 *Box pleats can be folded into any shape, like this "robot worm."*

Action origami figures that move make the most of paper's springy qualities. You may have seen the traditional hopping origami frog that leaps when you press it down and release it. This robotic-looking worm was inspired by Kinetogami robot worms being developed at Purdue University. Thanks to the box-pleat structure of this design, it will spring open when you press it flat and release it, like an inchworm that is trying to crawl along.

Materials

- Copy paper or origami paper
- Scissors (optional)

Figure 5-9 *To start, cut letter-sized paper into a square.*

Figure 5-10

Step 1

As with most origami, this model begins with a square sheet of paper. If you are using rectangular copy paper instead of square origami paper, first you must cut it into a square. To do this, lay the sheet of paper vertically on the table in front of you. Take the bottom-right corner and bring it up so that the bottom edge of the paper is even with the left edge. Fold down the top portion of the paper along the uppermost edge of the folded flap. Cut or tear off this top piece. Unfold the remaining square of paper. Fold it along the other diagonal and open it up. If you are starting with a square piece of paper, fold it diagonally in both directions.

Step 2

Take one corner, fold it up so that the corner touches the spot where the creases meet at the center of the paper, and unfold. Repeat with the other three corners:

Figure 5-11

Figure 5-12

Step 3

Now take one corner and fold it up so that the corner touches the crease you just made at the opposite corner where it crosses the original diagonal fold. Unfold. Repeat with the other three corners:

Figure 5-13

Step 4

Take one corner and fold it up so it touches the nearest crease and unfold. Repeat with the other three corners:

Figure 5-14

Step 5

Flip the paper over so the back is facing you. Fold the paper in half horizontally and unfold. Repeat vertically. Bring the bottom edge up even with the fold across the center of the paper and unfold. Do the same with the other three edges.

Figure 5-15

Step 6

Fold the bottom edge of the paper up even with the crease you just made on the top. Unfold and repeat with the other three edges.

Step 7

Fold the bottom edge up to the nearest crease and unfold. Do the same with the other three edges. You have now created the box pleats that you will use to build your robot worm.

Step 8

Fold up one of the corners of the sheet of paper along the nearest crease. Repeat with the other corners. Then fold the bottom edge up along the nearest crease. Repeat with the top edge:

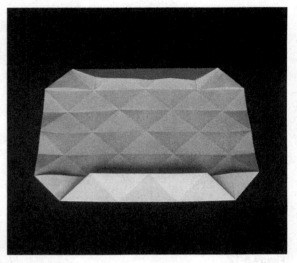

Figure 5-17

Step 9

Fold up the left edge of the sheet of paper along the nearest vertical crease:

Figure 5-16

Figure 5-18

Then fold it again along what is now the second-closest crease. Repeat with the right edge. The paper should look somewhat like a long, thin table turned upside down:

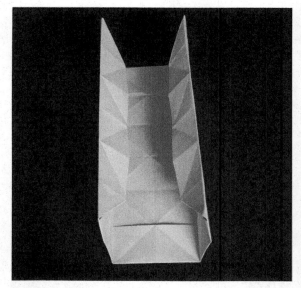

Figure 5-19

Step 10

Here's where it gets a little tricky. Pull up the middle of the flap along the bottom so the diagonal creases on either side form a triangle:

Figure 5-20

Tuck in the pleats that are created. Repeat with the top. Turn the model over (Figure 5-21).

Figure 5-21

The triangle you just made should fit next to triangles on either side.

Step 11

Across the top of the inchworm, you will see three large boxes with diagonals making an *X* inside each one. Begin to collapse the inchworm, accordion-style, by poking down the *X* in each box from the top while pressing in on the box's sides:

Figure 5-22

Keep pushing in the box pleats along the top. Take your time, as the first one may take some time to figure out. Let the mountain creases between each large box come up as the *X* goes down.

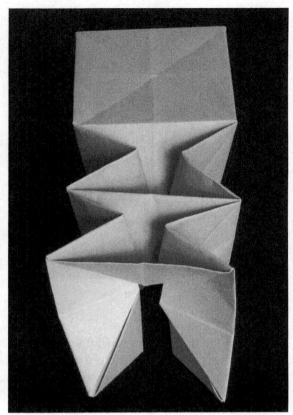

Figure 5-23

If you're having trouble, unfold the paper and make the creases a little sharper. When all the boxes are folded, press the accordion folds flat. Then stretch out the inchworm again.

Step 12

Turn the model so one leg of the "table" is facing you (Figure 5-24). Again you will see three large boxes with an *X* inside each one. In the first box, fold up the bottom of the *X* and pull it up until it sticks straight out, like the roof of a house:

Figure 5-24

Repeat with all the other boxes on both sides. When you are finished, fold the inchworm up accordion-style again to sharpen the creases a little more. Let it pop open. Your box-pleat paper robot inchworm is finished!

Figure 5-25 *The finished origami robot worm is made up of three collapsible segments.*

Figure 5-26 *Squeeze the worm flat, and it will pop open again as if it's trying to crawl.*

The Math of Cut Paper

Cutting paper is just as interesting to mathematicians as folding it. Sometimes, the two are combined, as in the fold and one cut problem solved by mathematician Erik Demaine. He proved that any two-dimensional shape with straight sides can be cut from a sheet of paper with only a single straight cut. The secret is folding the paper the right way before you cut it! To do that, you have to match up all the edges of the shape along one line that goes across the entire page when it is folded. Demaine has developed patterns for a swan, fish, butterfly and more.

Project: Fold-and-One-Cut Star

The most famous fold-and-cut shape is a five-pointed star. Legend has it that Betsy Ross used this method to make the stars on the first American flag, and a book by the famous magician Henry Houdini explains the trick. This version uses a letter-sized piece of paper, but there are ways to do the same trick with a square sheet of origami paper. See if you can figure them out!

Figure 5-27 *A star is a classic example of the fold-and-one-cut problem.*

Materials

- 8 1/2 by 11 inch paper
- Scissors

Step 1

Lay the sheet of paper down sideways. Fold it in half by taking the left edge and bringing it over to the right edge so that the shorter ends of the paper meet.

Step 2

Fold the upper-left corner of the paper down so it touches the middle of the bottom edge:

Figure 5-28

Step 3

Fold the lower-lefthand corner up along the slanted edge of the fold you just made:

Figure 5-29

Step 4

Fold the other top corner (the right edge) down over the new slanted edge you made:

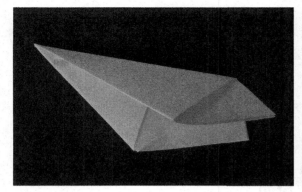

Figure 5-30

Step 5

Starting at the right folded edge, cut off the point on the left at an angle:

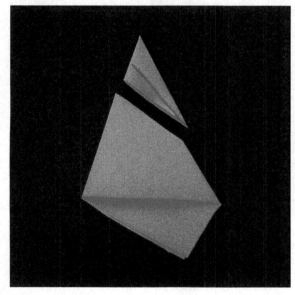

Figure 5-31

The lower you start the cut, the longer and thinner the arms of the star will be. Unfold the shape you cut out to see your star. When you unfold the outside of the cut, you get a star-shaped hole.

Figure 5-32 *Open up the top to reveal your star!*

For a fun variation, try snipping off the very tip of the point. Then cut across the point as you did the first time, to create several folded strips. As you unfold each one, you'll find you've got a series of stars in different sizes!

More Fold and One Cut

Do more of Erik Demaine's printable fold-and-one-cut patterns: *http://erikdemaine.org/foldcut/examples*.

Project: Make a Mobius Strip

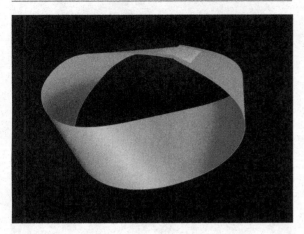

The Mobius strip is the granddaddy of cool paper math toys. Invented by the astronomer and mathematician August Ferdinand Möbius in 1858, the Mobius strip is simply a loop of paper with a twist in it. That twist gives it all sorts of interesting properties. For starters, a Mobius strip has only one side and one edge. And unusual things happen when you cut a Mobius strip in certain ways.

Mobius strips are fun to play with, but they're also useful in the real world. Manufacturers use them where they have loops that move along with their machinery, such as conveyor belts that pull objects along in an assembly line. That's because a Mobius strip will double the life of the loop by using both sides evenly. They're also used in the design of some electronic parts. And Mobius knitted scarves can be worn around your neck without having to tie the ends, but they're more pleasing to look at than a plain loop. Try making your own Mobius strip to see what amazing properties you can discover!

Materials

- Copy paper, adding machine paper, or other thin flexible paper

- Tape

- Scissors

Step 1

Cut a strip of paper long and thin enough to make into a flexible loop. For 11-inch-long paper, a width of 1 1/2 inches will work fine.

Step 2

Put a piece of tape hanging off one end of the strip, or work with a partner to hold and tape your loop.

Step 3

Twist one end of the strip so that the back is facing you:

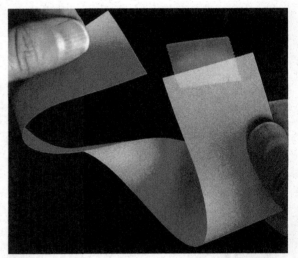

Figure 5-33

Then tape the ends of the paper together to make a loop. That's it! What can you do with your Mobius strip? Here are some ideas:

Step 4

Take a marker and draw a line down the middle. Notice that you have to go around the loop twice before you reach the place you started.

Step 5

Cut the Mobius strip along the line you just drew:

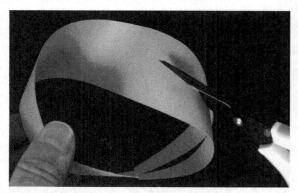

Figure 5-34

If this were a normal loop, you'd end up with two loops. What happens to the Mobius strip is a little different.

Step 6

Cut the result of the step above down the middle again. Now what do you have?

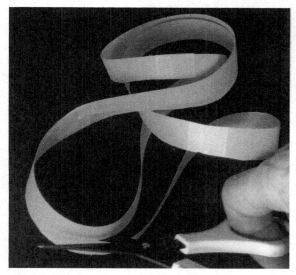

Figure 5-35

Step 7

Make a new Mobius strip and try cutting it about a third of the way in. What happens this time?

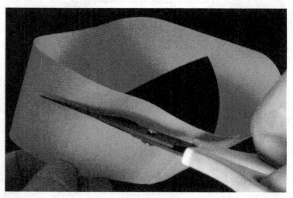

Figure 5-36

Step 8

Try adding extra twists to your strip as you make it. For instance, make a strip with three twists (technically half-twists since you're only turning the paper front to back, not all the way around). Cut it down the middle as before, and you will get a completely different result.

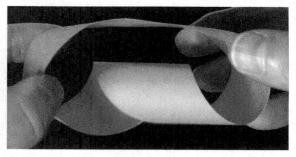

Figure 5-37 *What happens when you cut apart a Mobius strip with three twists?*

(To see if you got the expected result with each experiment, see the photos in Appendix E!)

Project: Make a Hexaflexagon

Figure 5-38 *For such a simple design, a hexaflexagon has amazing powers.*

The hexaflexagon is a favorite folded-paper math toy because it's so much fun to flip it around to make it reveal its hidden wonders. It was invented in 1939 by a bored math student, Arthur Stone, who was playing around with some strips of paper in class. But the hexaflexagon didn't become widely known outside math departments until a magician named Martin Gardner wrote an article about it for *Scientific American* in 1956. Gardner went on to write a regular column and many books about recreational mathematics that inspired generations of math fans. Vi Hart made a series of videos on YouTube about hexaflexagons that made them truly famous. Here is how to make a three-sided, or trihexaflexagon, based on Vi Hart's videos.

Materials

- Paper (if you have any adding machine paper from earlier projects, that will work here, too!)
- Scissors
- Tape

- Crayons or colored pencils or markers that will not bleed through the paper (such as highlighters)

 Hexaflexagon Tips

To make it easier to tell what's happening, use one-sided origami paper, as in the photographs, or fill in the front and back of the hexaflexagon using different colors.

If at any point your hexaflexagon doesn't open easily, like the petals of a flower, try pulling the opposite creases out (make the valley folds into mountain folds).

The Mobius-Hexaflexagon Connection

Strangely enough, a trihexaflexagon is basically a three-twist Mobius strip. It has one side and one edge. The main difference is that hexaflexagon has been squashed flat!

Step 1

Take a long, narrow strip of paper. If you are using paper that is 11 inches long, cut your strip 1 1/2 inches wide. Hold one end of the strip in your left hand and let it hang down. Take the bottom of the strip in your right hand. Bring it back and over to the right, making a loose curvy *L* shape. The paper will be twisted so the back is facing you on the right (Figure 5-39).

Figure 5-40

Step 3

Bring the right end of the strip in front of the left end. The paper is now in the shape of a six-sided hexagon (you'll fix the top of the hexagon in a minute):

Figure 5-39 *Bend the strip of paper into an L. Origami paper with color on one side is used here to make it easier to tell the front from the back.*

Step 2

Twist the right end of the strip again in the same direction so that the front of the paper is facing you. Then bring the right end back and up, making a loose curvy *U* shape:

Figure 5-41

Adjust the place where the two ends cross so the gap in the middle of the shape is nearly closed. Try to make the five bottom sides equal.

Step 4

Put the hexaflexagon down on your work surface and press it flat, creasing the folds sharply:

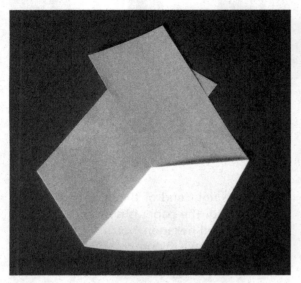

Figure 5-42

Use scissors to cut straight across the place where the two ends cross:

Figure 5-43

Put some tape over the ends you just cut to hold them together (Figure 5-1).

Step 5

Fold the hexaflexagon shape you just created in half so that two opposite straight edges, not the points, meet:

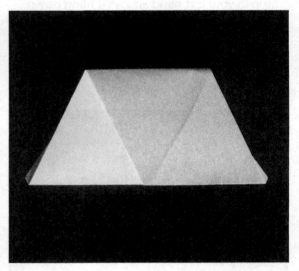

Figure 5-44

Unfold it, then fold it along the same crease, but in the other direction. In other words, if your first fold was a valley fold, your next should be a mountain fold. Unfold the hexaflexagon and do the same to the other two straight edges.

Step 6

Finally, fold along all the creases until you have made one small but fat triangle:

Figure 5-45

The open it up and pinch it to make three triangular "wings" stick out that can be turned like the pages of a book:

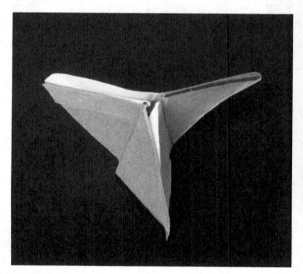

Figure 5-46

The hexaflexagon begins to open like the petals of a flower!

Figure 5-47

Carefully pull the wings apart, starting at the top, until your hexaflexagon is flat again. Open it up and a new side appears:

Figure 5-48

Step 7

Which side is facing up now? Is it the first, or the second? If you did everything right, you should be looking at a third side you haven't seen before!

Figure 5-49

Repeat step 6. What happens as you continue to cycle through the different sides? Is there any end to the rotation, or can you keep going in that direction forever? Can you reverse directions and go the other way?

You can get some interesting effects by drawing designs on your hexaflexagon. Flip the hexaflexagon over and your drawings will turn inside out! If you'd like to make a hexaflexagon with even more sides, check out the video by Vi Hart and the other resources below.

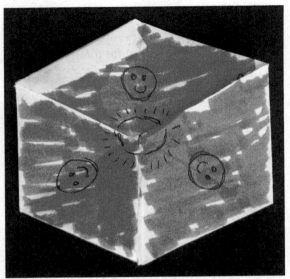

Figure 5-50 *Color in your hexaflexagon and add some drawings to see more unusual effects. This one starts with a green side.*

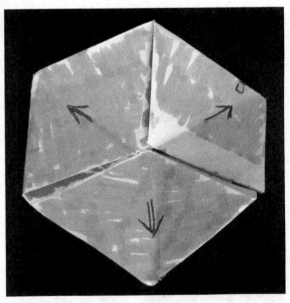

Figure 5-51 *Next comes a pink side.*

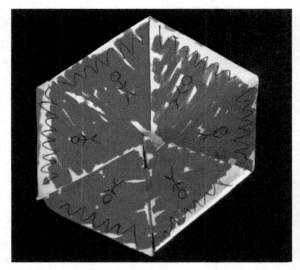

Figure 5-52 *Then an orange side.*

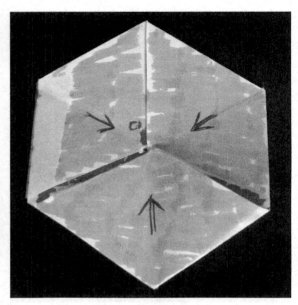

Figure 5-54 *These pink arrows were pointing the other way on the front.*

Figure 5-53 *Flip it over, and the green side drawings turn inside out.*

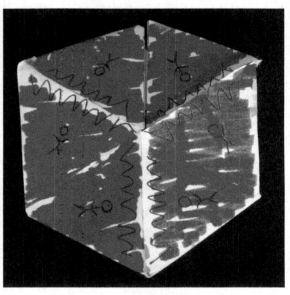

Figure 5-55 *And these orange people are now separated by wiggly lines.*

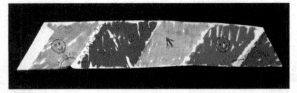

Figure 5-56 *Open up your hexaflexagon to see how it works.*

Figure 5-57 *The back of the hexaflexagon strip.*

More About Hexaflexagons

- Vi Hart's hexaflexagon videos on Khan Academy (*http://khanacademy.org/math/recreational-math/vi-hart/hexa flexagons/v/hexaflexagons*)

- Martin Gardner's article (*http://www.maa.org/publications/periodicals/college-mathematics-journal/college-mathematics-journal-contents-janu ary-7*) on hexaflexagons, reprinted in *The College Mathematics Journal*

- Flexagon.net (*http://flexagon.net*)

Templates

Tips for Copying or Drawing Your Own Templates

These templates are copies of the ones you'll see throughout the chapters, gathered here to make them easier to copy by machine or hand. The original sizes are indicated where helpful, but they can all be modified to fit the kind of project you want to build. Just make sure that parts that need to fit together (like the rods that must fit through the guides on the paper machines) still fit together if you resize them.

Self-Folding Paper Model Templates

Figure A-1 *Paper pyramid*

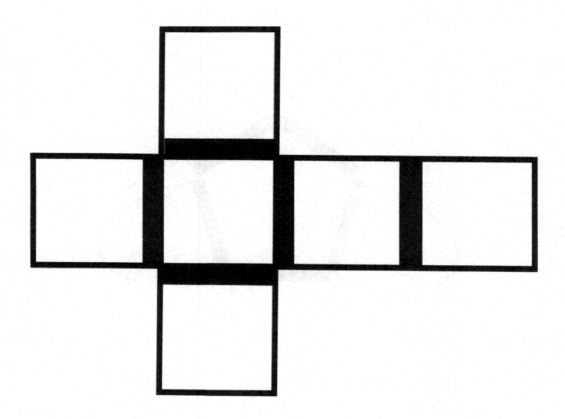

Figure A-2 *Cube template*

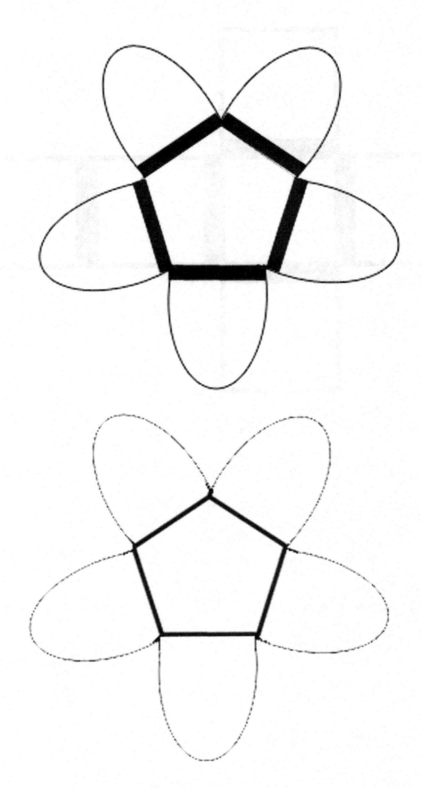

Figure A-3 *Flower template*

Paper Tech Templates

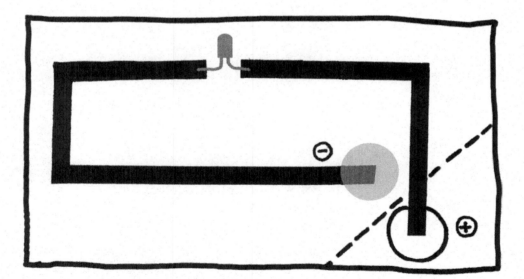

Figure B-1 *Simple circuit template*

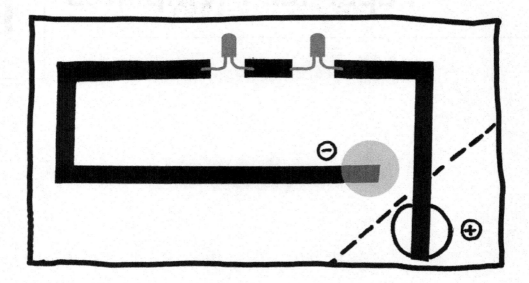

Figure B-2 *Two LEDs in series*

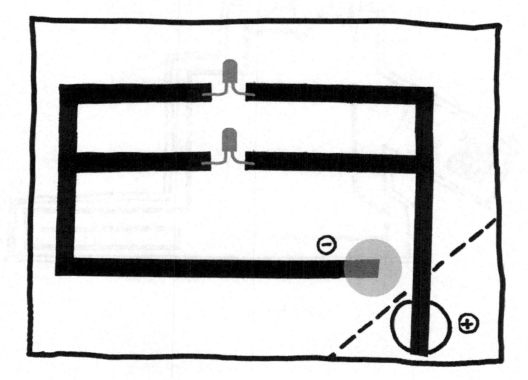

Figure B-3 *Two LEDs in parallel*

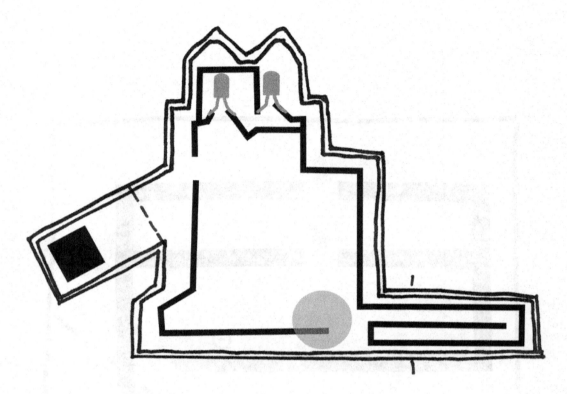

Figure B-4 *Light-Up Cat*

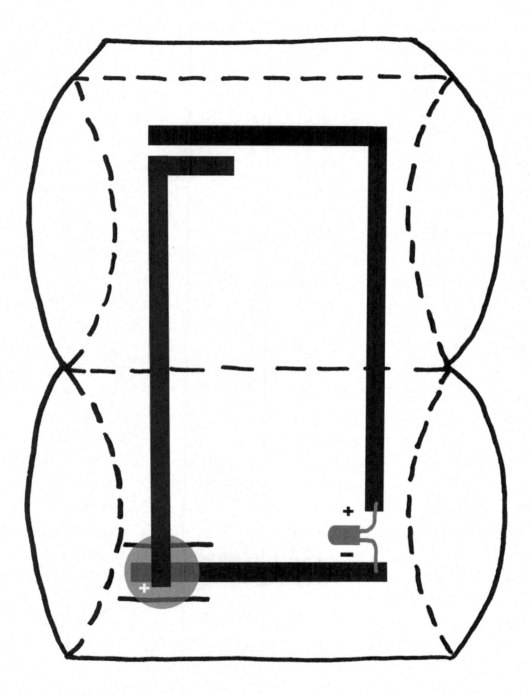

Figure B-5 *Tilt sensor*

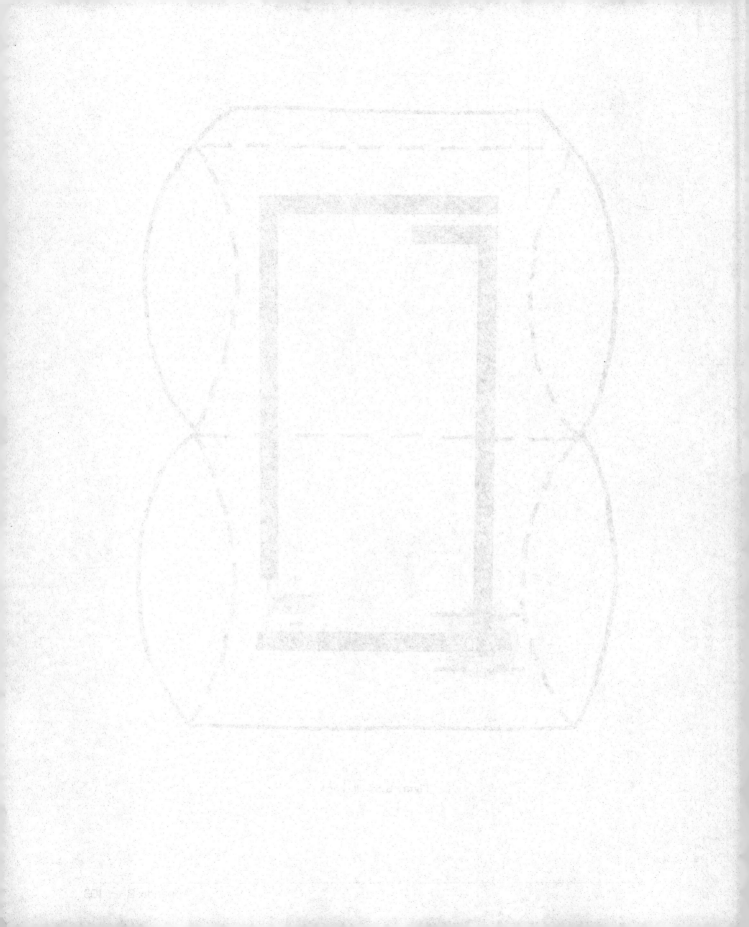

Cam Sampler

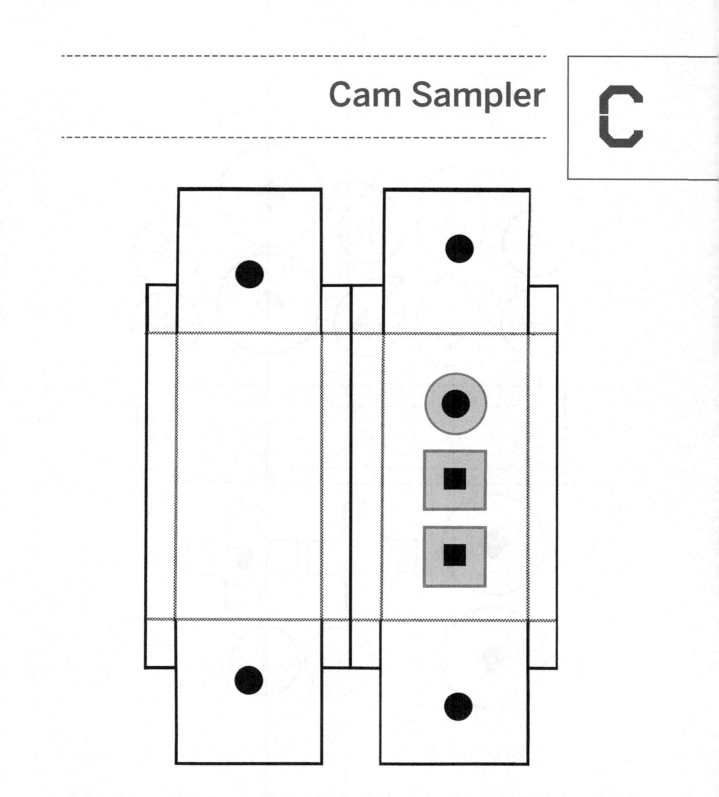

Figure C-1 *When folded into shape, the Cam Sampler base is 5 inches long, 3 inches wide, and 3 inches high. The side flaps are 1/2 inch deep.*

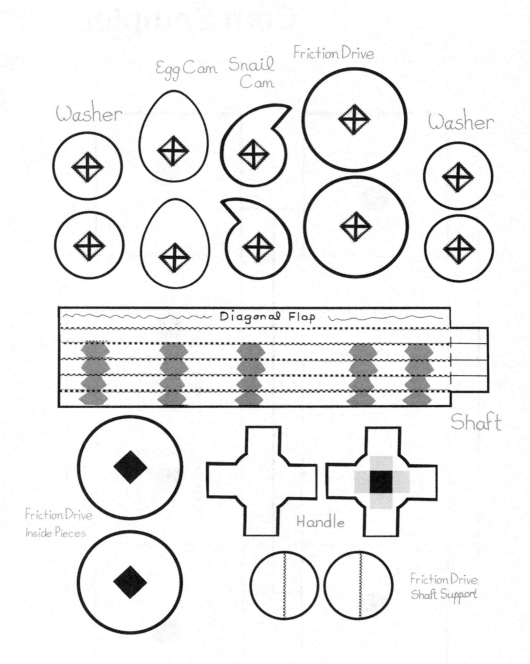

Figure C-2 *When folded into shape, the shaft is 6 inches long. It forms a hollow square measuring 1/4 inch on each side. All the small circle shapes are 1 inch across. The large circles are about 1 1/2 inches across.*

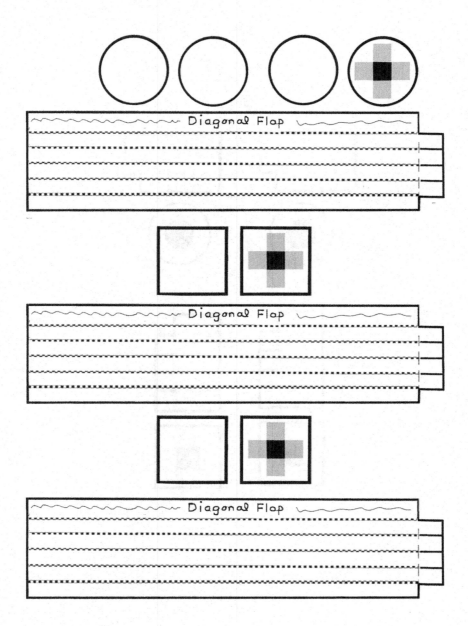

Figure C-3 *When folded into shape, the rods are 6 inches long, 1/4 inch wide, and 1/4 inch high. The tabs are 1/4 inch long.*

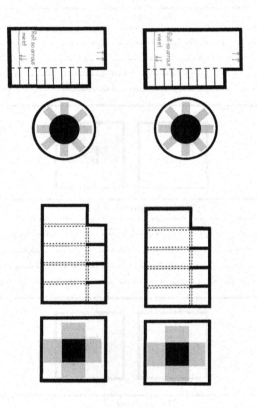

Figure C-4 *When folded into shape, the square guides are a little less than 3/8 inches on a side. Each side of the square base is about 1 inch. The base of the round guide is also 1 inch across.*

Space Rover

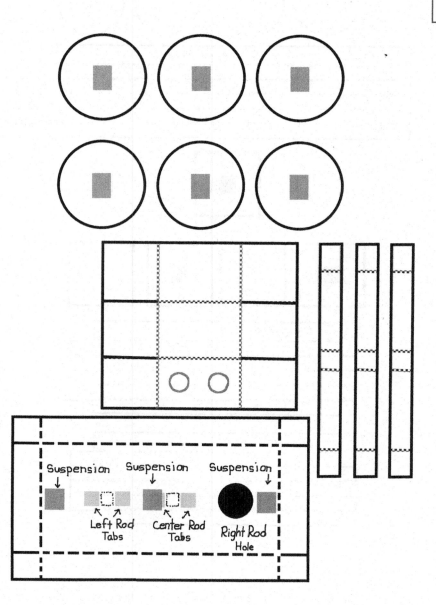

Figure D-1 *When folded into shape, the rover body is 4 3/8 inches long and 2 inches wide. The side flaps are 1/2 inch deep. The head is 1 1/2 inches long, 1 inch wide, and 1 inch high. The suspensions are 4 1/4 inches long (unfolded) and 5/8 inches wide. The wheels are about 1 1/8 inches across.*

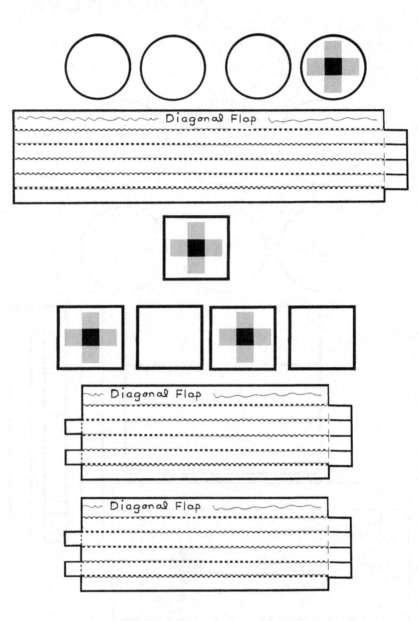

Figure D-2 *When folded into shape, the top rod is 6 inches long and the bottom rods are 4 inches long. All are 1/4 inch wide and high.*

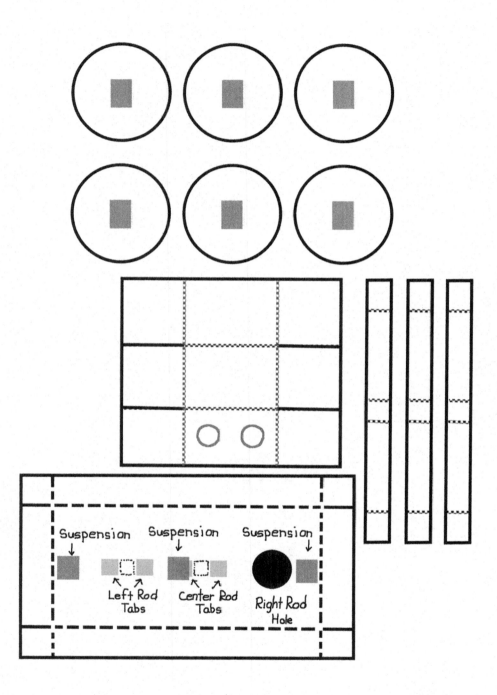

Figure D-3 *When folded into shape, the rover body is 4 3/8 inches long and 2 inches wide. The side flaps are 1/2 inch deep. The head is 1 1/2 inches long, 1 inch wide, and 1 inch high. The suspensions are 4 1/4 inches long (unfolded) and 5/8 inches wide. The wheels are about 1 1/8 inches across.*

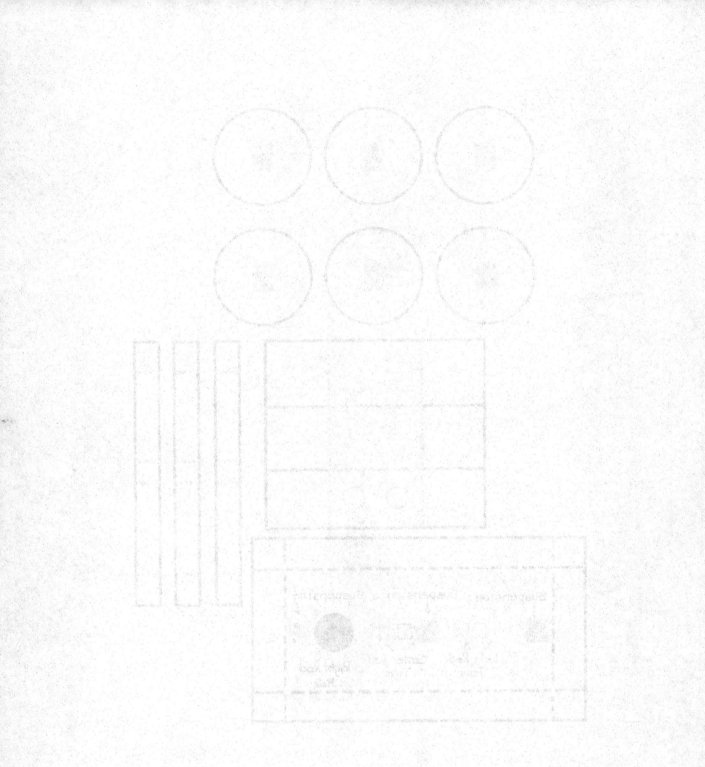

Mobius Strip Results

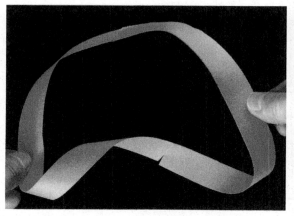

Figure E-1 *One big loop is what you get when you cut a Mobius strip in half!*

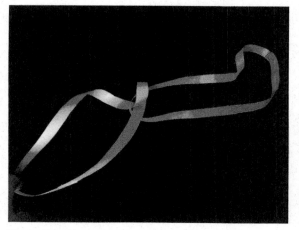

Figure E-2 *Two twisted loops are created when you cut the above loop in half.*

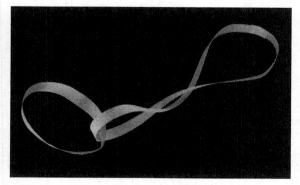

Figure E-3 *Cut a Mobius strip in thirds, and you get two linked loops!*

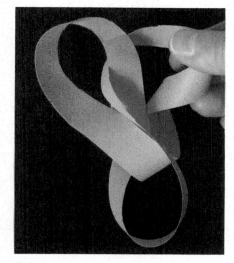

Figure E-4 *Cut a triple-twist Mobius strip in half, and you get one long loop in a pretzel-shaped trefoil knot!*

About the Author

Kathy Ceceri is the author of fun and educational books for kids and families, including *Making Simple Robots: Explore Cutting Edge Robotics with Everyday Stuff*, *Robotics: Discover the Science and Technology of the Future*, and *Video Games: Design and Code Your Own Adventure*. She helped create the GeekMom blog and the book *Geek Mom: Projects, Tips, and Adventures for Moms and Their 21st-Century Families* and contributed more than a dozen projects to the Geek Dad series of books. Formerly the homeschooling expert at About.com, Kathy presents STEAM workshops at schools, museums, libraries, and maker faires around the country. She lives with her family in upstate New York, and her website is *craftsforlearning.com*.

Colophon

The cover and body font is Benton Sans, and the heading font is Serifa.

CPSIA information can be obtained at www.ICGtesting.com
Printed in the USA
BVOW10n1047121015

421776BV00001B/1/P